MW00844793

Wastewater Biology: The Microlife

A Special Publication

Second Edition

Prepared by **Wastewater Biology: The Microlife** Task Force of the **Water Environment Federation**

Ronald G. Schuyler, *Chair*

Thomas R. Barron	Robert P. Marino
Jerry T. Cheshuk	Mary Alice H. Nelson
Ernst M. Davis	Mesut Sezgin
Joseph J. Gauthier	Sophie G. Simon
Edwin E. Geldreich	Thomas L. Stokes, Jr.
Roger R. Hlavek	Alan Warren
Tracy L. Finnegan	Richard G. Weigand
Dennis R. Lindeke	Robert A. Whitworth
Charles L. Logue	Melvin C. Zimmerman

Under the Direction of the **Municipal Subcommittee** of the **Technical Practice Committee**

2001

Water Environment Federation
601 Wythe Street
Alexandria, VA 22314–1994 USA

IMPORTANT NOTICE

The material presented in this publication has been prepared in accordance with generally recognized engineering principles and practices and is for general information only. This information should not be used without first securing competent advice with respect to its suitability for any general or specific application.

The contents of this publication are not intended to be a standard of the Water Environment Federation (WEF) and are not intended for use as a reference in purchase specifications, contracts, regulations, statutes, or any other legal document.

No reference made in this publication to any specific method, product, process, or service constitutes or implies an endorsement, recommendation, or warranty thereof by WEF.

WEF makes no representation or warranty of any kind, whether expressed or implied, concerning the accuracy, product, or process discussed in this publication and assumes no liability.

Anyone using this information assumes all liability arising from such use, including but not limited to infringement of any patent or patents.

Library of Congress Cataloging-in-Publication Data

Wastewater biology. The microlife: a special publication / prepared by Wastewater Biology: the Microlife Task Force of the Water Environment Federation; under the direction of the Municipal Subcommittee of the Technical Practice Committee.—2nd ed.
    p. cm.
Includes bibliographical references and index.
ISBN 978-1-57278-167-2
1. Sewage—Microbiology. I. Water Environment Federation. Wastewater Biology: the Microlife Task Force. II. Water Environment Federation. Municipal Subcommittee.

TD736.W37 2001
628'.3'01578—dc21   2001045543

Water Environment Federation

Founded in 1928, the Water Environment Federation (WEF) is a not-for-profit technical and educational organization with members from varied disciplines who work toward the WEF vision of preservation and enhancement of the global water environment. The WEF network includes more than 100,000 water quality professionals from 77 Member Associations in 31 countries.

For information on membership, publications, and conferences, contact

Water Environment Federation
601 Wythe Street
Alexandria, VA 22314-1994 USA
(703) 684-2400
http://www.wef.org

Special Publications
of the Water Environment Federation

The WEF Technical Practice Committee (formerly the Committee on Sewage and Industrial Wastes Practice of the Federation of Sewage and Industrial Wastes Associations) was created by the Federation Board of Control on October 11, 1941. The primary function of the Committee is to originate and produce, through appropriate subcommittees, special publications dealing with technical aspects of the broad interests of the Federation. These publications are intended to provide background information through a review of technical practices and detailed procedures that research and experience have shown to be functional and practical.

Contents

List of Tables

List of Figures

List of Plates

Preface

The description, ecology, and beneficial and detrimental roles of microscopic life forms—the microlife—found in wastewater treatment processes are critical biological knowledge for operators and technicians. Operators and technicians must regulate the wastewater environment to provide acceptable treatment conditions. This publication emphasizes the biological aspects of wastewater treatment that would be of most value to them. This book provides a general overview of microorganisms and is written with a minimum of technical jargon so that it can be of value to a wide spectrum of wastewater professionals.

Because the manual includes discussions of the taxonomy (the naming and classification) of the microlife commonly found in a wastewater treatment plant, a brief review of taxonomic principles is included. The primary means of identifying the microlife is through the microscope, therefore, a chapter is dedicated to its use and understanding. In addition to the microscope, a means of keeping records of the microlife can be obtained through photomicroscopy, the technique of taking photographs through the microscope. Therefore, a chapter is also dedicated to photomicroscopy.

This publication provides operators and technicians, as well as other wastewater professionals, with an easy-to-use guide to the biological aspects of wastewater treatment, including 13 color plates to assist in identifying the microlife.

Acknowledgments

Principal authors of the publication are

Chapter One	Ronald G. Schuyler
Chapter Two	Richard G. Weigand
Chapter Three	Melvin C. Zimmerman
Chapter Four	Robert P. Marino
Chapter Five	Alan Warren and Colin Curds
Chapter Six	Robert A. Whitworth
Chapter Seven	Mary Alice H. Nelson and Ronald G. Schuyler
Chapter Eight	Mesut Sezgin and Joseph J. Gauthier
Chapter Nine	Robert P. Marino
Chapter Ten	Edwin E. Geldreich
Chapter Eleven	Melvin C. Zimmerman
Chapter Twelve	Tracy L. Finnegan

Authors' and reviewers' efforts were supported by the following organizations:

City of Atlanta Public Works Department, Atlanta, Georgia
Lycoming College, Williamsport, Pennsylvania
McCombs Frank Roos Associates, Inc., Plymouth, Minnesota
Nalco Chemical Company, Naperville, Illinois
The Natural History Museum, London, United Kingdom
Rothburg, Tamburini and Winsor, Inc., Denver, Colorado
University of Alabama, Birmingham
West Virginia Environmental Training Center, Ripley, West Virginia

Chapter 1
Introduction

OVERVIEW

Wastewater Biology: The Microlife is a presentation of the description, ecology, and beneficial and detrimental roles of the microscopic life forms—the microlife—found in wastewater treatment processes. Because a knowledge of biology is especially desirable for the operators, technicians, and engineers who must regulate and design the wastewater environment to provide acceptable treatment conditions, this manual emphasizes the biological aspects of wastewater treatment that would be of the most value to them. However, the text also will be of value to other wastewater professionals, especially chemists, educators, and microbiologists.

This publication has been written with a minimum of technical language so that it can be easily reviewed and absorbed by individuals from different professional backgrounds. However, because the manual includes discussions of the taxonomy (naming and classification) of the microlife typically found in a wastewater treatment plant, a brief review of taxonomy principles is included here.

TAXONOMY

To deal with the enormous diversity of life forms, organisms historically have been grouped by one or more related characteristics they display (such as structure, genetics, or biochemistry) into a hierarchy of categories: kingdom, phylum, class, order, family, genus, and species. The top category, kingdom, is a very broad, general category. As more categories are specified, the number of different organisms that fit that particular classification decreases. When the last category, species, is specified, only one life form or organism is designated.

As the number of known life forms or species has increased along with our knowledge of the degrees of relationships of these species, the need for a more precise indication of the classification position of species has also occurred. To cope with this need, classifiers have split the original seven basic categories and inserted additional categories. Modern classification schemes, therefore, consist of the following: kingdom, phylum, subphylum, superclass, class, subclass, cohort, superorder, order, suborder, superfamily, family, subfamily, tribe, genus, subgenus, species, and subspecies.

It is not necessary, however, to list all categories when discussing a particular organism. The scientific name (bionomen) consists of the generic name (denoting the genus) and the species name that distinguishes the organism from other species of the genus. For example, the honeybee is *Apis mellifera*. Note that the genus is capitalized, but the species name is not. Also, both genus and species names are either underlined or set in italic type. If the scientific name of the organism is repeated in text, the genus name is abbreviated, for example, *A. mellifera*.

USE OF THE MANUAL

The groups of life forms reviewed in this manual include the organisms considered the most important in the treatment of wastewater and disease transmission. Understanding these microorganisms will assist the operator, engineer, and others in better understanding how to properly control processes. Other organisms, such as viruses that have little or no treatment significance, are not included in this manual, except as they relate to water-borne diseases (Chapter 10).

Chapter 2 gives an overview of the microorganisms. Chapter 3 then introduces a valuable tool in the observation of microorganisms, the microscope. Discussions of the specific types of organisms begins in Chapter 4 with the bacteria and continues with the protozoa (Chapter 5), rotifers (Chapter 6), nematodes and other metazoa (Chapter 7), and filamentous organisms in Chapter 8. These are the organisms typically involved in wastewater treatment. However, some of these organisms are also pathogenic or parasitic and cause disease. Chapter 9 discusses bacteria that can be used to indicate the potential presence of disease-causing microorganisms. Those disease-causing organisms, or pathogens, are reviewed in Chapter 10. Chapter 11 completes the discussion of microorganisms by covering the parasites, those organisms that live at the expense of host organism. Chapter 12 presents information on equipment and techniques to obtain a photographic record of microscopic findings. The equipment and procedures necessary for examining, identifying, and enumerating these life forms also are presented in the Appendix. Hopefully, with this content and arrangement, readers will find *Wastewater Biology: The Microlife* useful in their professional endeavors and a source of information whenever questions arise relative to microorganisms.

Chapter 2
The Microorganisms

HISTORY

BEFORE MICROSCOPES. Robert Hooke, in the 1660s, is credited with assembling lenses into the first crude microscope (Figure 2.1). However, another craftsman, Antonj Van Leeuwenhoek, was the first to actually study and describe bacteria and protozoa in great detail, terming them *animalcules*. Using simple but powerful lenses of approximately 300X magnification, he studied a variety of objects, including cow dung.

Leeuwenhoek's invention gave scientists the power to see, for the first time, an incredible new universe of tiny objects so minute that their existence had never been suspected. This was a momentous discovery that had many applications such as understanding food spoilage, fermentation, and the causes of human disease.

GERM THEORY OF DISEASE. After Hooke's original invention of the microscope, it took more than 150 years to develop instruments even remotely similar to ones in use today. By that time scientists were beginning to understand the relationship between certain diseases and their microscopic causes.

In the 1880s, Robert Koch established the bacterial cause of cholera and developed a series of steps or postulates to link specific microbes to specific

Figure 2.1 Hooke's compound microscope, drawn by himself. Note that
the object is seen by light reflected from above, unlike
Leeuwenhoek's "animalcules," which were seen by light
transmitted through them.

diseases. Pathogenic, or disease-causing microorganisms, were now being
discovered.

French scientist Louis Pasteur is credited with advancing much of this
knowledge as a result of his research on why German beer was considered
superior to French brew. His experiments included heating, or pasteurization,
of liquids to kill unwanted organisms, an idea that was later applied to the
medical field. The general understanding of germs grew into specific knowledge of the relationship between one disease and one type of microorganism.
Soon these specific causative agents were isolated and identified.

In the late 1800s, this led to the realization that epidemics such as cholera
and typhoid were waterborne diseases of microbiological origin. Suddenly
drinking water treatment and wastewater treatment were critical industries and
the realm of pollution microbiology was born.

TYPES OF MICROORGANISMS

Microorganisms of greatest significance to water quality professionals can be classified simply into four groups: bacteria, protozoa, metazoa, and viruses. It should be understood that the descriptions provided here will serve as working definitions for these groups, with detailed descriptions to be found elsewhere. Each of these groups plays a key role in the complex world of wastewater biology.

BACTERIA. These small, single-celled organisms are found in tremendous numbers in polluted water. They come in a variety of sizes and shapes and are the most prolific microorganisms on earth, found essentially everywhere.

Many significant water-borne diseases have bacteria as their causative agents and bacteria do the most significant portion of the work in wastewater treatment. Of particular interest to wastewater operators are those bacteria that grow as long, threadlike filaments and those that help form small biological communities known as floc (Figure 2.2). Chapter 4 of this manual presents a more detailed discussion of bacteria.

PROTOZOA. This term applies to a rather large and diverse group of single-cell organisms that are typically much larger than bacteria. Protozoa, like bacteria, are typically found in large numbers in polluted water and waste-water. Many of them are quite mobile, swimming rapidly through their

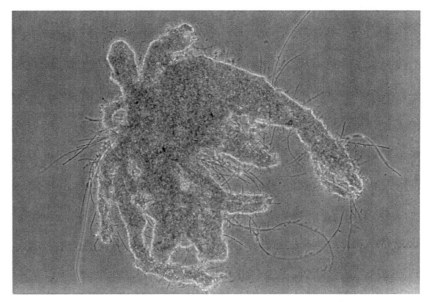

Figure 2.2   Individual bacteria grow as a colony to form a floc particle (courtesy of Richard G. Weigand).

---

microscopic world. Others attach themselves to tiny particles and extend into the water around them (Figure 2.3).

Protozoa play a critical role in the wastewater treatment process. Their presence in varying numbers has long been used as a process evaluation tool for experienced operators. Some protozoa play primary roles in the clarification of effluent because of their predation on dispersed bacterial growth. In many instances specific protozoa can be easily identified and quantified in wastewater samples.

During the past decade, protozoans have received increased attention because of their connection with waterborne disease. *Giardia* and *Cryptosporidium* are two protozoans responsible for significant outbreaks of illness in the United States. Chapter 5 of this manual describes the protozoa in more detail.

METAZOA. This term technically applies to all animals composed of more than one cell. From a practical standpoint it refers to larger, more complex animals than those considered protozoa. Included in the metazoa group are organisms known as rotifers, so named because of their rotating crowns of hairlike cilia (Figure 2.4). Although typically somewhat larger than bacteria and protozoa, they are still microscopic in size.

Rotifers are fairly common in some wastewater treatment processes and play an active role in the breakdown of organic wastes. Observing and quantifying rotifers contributes to the evaluation of treatment conditions. Chapter 6 of this manual is devoted to further discussion of rotifers.

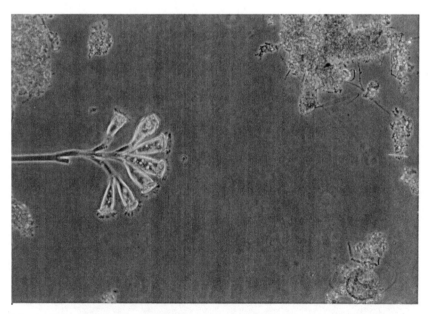

Figure 2.3   Some protozoa such as these stalked ciliates attach to floc particles and extend out into the water around them (courtesy of Richard G. Weigand).

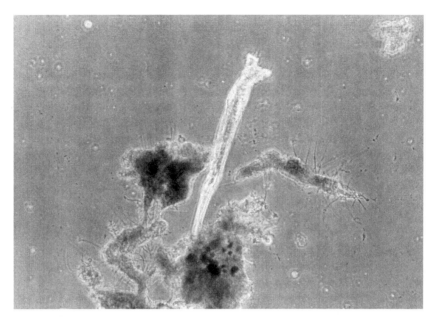

Figure 2.4   A typical rotifer, commonly found in wastewater treatment systems (courtesy of Richard G. Weigand).

Many other metazoa must be considered in the discussion of wastewater microbiology. Nematodes, or roundworms, have many microscopic forms often found in wastewater treatment. Water bears, bristle worms, water fleas, and seed shrimp are all metazoa that may take up residence in a wastewater treatment process. These higher life forms are the most complex microorganisms typically associated with wastewater microbiology and are further discussed in Chapter 7.

VIRUSES. This group of organisms is somewhat unique compared with those described above. They are much smaller than bacteria, too small to be seen with a conventional microscope. Viruses are parasites and therefore can only reproduce within a host cell (such as a bacteria, animal, or plant cell.).

Although viruses play no known role in wastewater treatment processes, they are nonetheless important to water quality professionals. Many diseases caused by viruses can be transmitted through polluted waters. Hepatitis is the most common one. However, the human immunodeficiency virus, the virus that causes acquired immune deficiency syndrome, has never been shown to be transmitted through wastewater (WEF, 1991 and 1999).

More information on viruses can be found in Chapter 10 on Wastewater Pathogens.

ROLES OF MICROORGANISMS

PARASITES AND PATHOGENS. A wide range of pathogenetic organisms can be found in wastewater samples. These may include members of all the groups described above and worms and fungi. These pathogens can be transmitted to the wastewater professional through contact, accidental ingestion, or inhalation. For this reason, operations personnel should take appropriate precautions and use protective equipment such as rubber gloves, face shields, aprons, and safety glasses. Such safety measures are especially important when dealing with raw wastewater (WEF, 1994). Many cases of illness could possibly be traced to careless work practices.

SOIL MICROBES AND HETEROTROPHS. Microorganisms are important to both the formation and fertility of soil. Bacteria, protozoans, and nematodes are all found in typical soils in large numbers. Many of these microbes find their way into wastewater collection systems and ultimately contribute to microlife found in various treatment processes. Natural chemical cycles such as the carbon cycle, nitrogen cycle, and sulfur cycle rely heavily on some of the bacteria found in soils. These cycles have a profound effect on the performance of all treatment units.

The term heterotroph refers to any organism that uses some form of organic material as an energy source. Protozoa, metazoa, and most bacteria are heterotrophs. The wastewater treatment process relies on these heterotrophic microorganisms to eat tremendous quantities of organic material in the wastewater. Specific treatment units are intended to enhance the growth of these organisms so that dissolved organic wastes can be converted to additional biomass that can be more easily removed from the waste stream.

FRESHWATER MICROBES. The numbers and kinds of microorganisms in natural water bodies depend largely on the available nutrient supply, environment, and other organisms present. Of primary importance are the types and quantities of nutrients and natural levels of available oxygen in the water. Even the most pristine mountain stream can contain organic or chemical nutrients capable of supporting some forms of bacteria.

Typically, lakes and streams contain organic material from aquatic life, animals, and plants in quantities sufficient to support a wide range of microorganisms. All of these organisms form a complex food web that gradually stabilizes their environment.

ECOLOGICAL RELATIONSHIPS. Of vital importance in the ecological cycles of the earth are the hundreds of common species of microorganisms, found both in soil and water, that decompose the organic material in wastes. Their byproducts are then recycled as soluble and relatively inoffensive or potentially useful materials (such as methane). Technologies exploiting microorganisms as agents for the prevention of pollution have been used for decades with great success. Under current conditions of human activity almost no significant water body exists long without some form of microbial presence.

The wastewater treatment process is simply a dramatically compressed and somewhat controllable version of nature's own purification process. The primary objective of the plant operator and design engineer is to optimize conditions for the actual treatment providers, Leeuwenhoek's animalcules.

*R*EFERENCES

Water Environment Federation (1991) *Biological Hazards at Wastewater Treatment Facilities*. Special Publication, Alexandria, Va.

Water Environment Federation (1994) *Safety and Health in Wastewater Systems*. Manual of Practice No. 1, Alexandria, Va.

Water Environment Federation (1999) *HIV in Wastewater: Presence, Survivability, and Risk to Wastewater Treatment Plant Workers*. Alexandria, Va.

Chapter 3
The Microscope

*I*NTRODUCTION

Microscopes are used routinely to identify and enumerate (count) many different types of organisms present in wastewater. This manual discusses a variety of microscope applications for examining wastewater organisms. The purpose of this chapter is to provide an overview of microscopy.

Two types of microscopes are typically used: a high-magnification micro-scope composed of a one compound-lens system and a stereoscopic binocular microscope with two complete-lens systems for producing an image with significantly lower magnification. The term compound microscope refers to the high-magnification microscopes that provide total magnification from 40 to 1000X. Traditionally, these microscopes had a singular ocular lens (or eyepiece) and were called monocular. Currently, most compound microscopes are binocular and have two ocular lenses that help reduce eyestrain but do not provide stereoscopic vision. The stereoscopic binocular microscope, also referred to as a dissection microscope, provides lower magnification from 8 to 40X. Because they are composed of two complete-lens systems (one for each eye), these microscopes allow the user true stereoscopic vision. Figure 3.1 compares the monocular compound and stereoscopic binocular microscopes.

COMPOUND MICROSCOPE

DESCRIPTION. All compound microscopes contain a basic-lens system consisting of an objective lens that produces an image, typically magnified, of the object observed and an ocular lens that further magnifies the image. The objective and ocular lenses are situated at opposite ends of the microscope's body tube with the objective located closer to the specimen. The total magni-fication of a given compound-lens system is the product of the magnification of the ocular lens (typically 10X) and the magnification of the objective lens (which can be varied).

The number of objective lenses on a microscope typically varies from one to four and are held in place on a revolving nosepiece. A typical combination for magnification on a microscope with four objective lenses is 4, 10, 40 and 100X. The magnification of a particular lens is noted on the side of its barrel. Most often the 100X objective is an oil-immersion lens. If so marked, immersion oil is to be used with this lens with its front lens in contact with the cover glass.

Focusing a microscope is accomplished by changing the position of the body tube with respect to the position of the stage, the platform on which the specimen to be observed is placed. In most modern microscopes, the stage is movable. Illumination of the specimen comes from a light beneath the stage. Proper illumination is extremely important because improper lighting will produce a poor image, inaccurate observations, or unnecessary eyestrain. Most microscopes have built-in light sources, but others have mirrors beneath the stage and illumination is provided by separate lamps.

For most applications, special lamps that provide blue, diffuse light should be used. The lamp should be positioned in front of the microscope and the substage mirror adjusted to obtain an evenly lit field of view. Most substage mirrors have two surfaces: a flat one and a concave one. The flat mirror is the better one to use if illumination is provided by a nearby lamp. The concave mirror is typically used only when there is a need to concentrate the light rays when the microscope has no condenser.

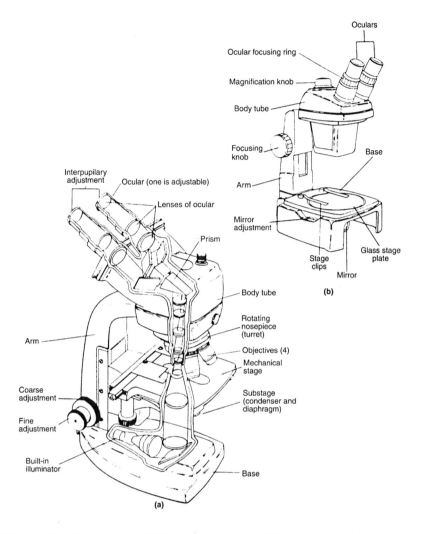

Figure 3.1 Comparison of (a) compound and (b) stereoscopic binocular microscopes. The compound microscope is sectioned to show pathway of light from illuminator to eyepiece.

Below the stage, a microscope has one or more of the following devices for regulating the intensity of light passing through the specimen:

- Rotatable disc diaphragm for selecting the appropriate size hole through which light must pass,
- Condenser with an iris diaphragm with the aperture size regulated by an adjustable lever (Glase et al., 1979), and
- Built-in lamp with a rheostat to regulate electrical current and, consequently, light intensity.

For those microscopes that do not have rheostats, adjusting the iris diaphragm is typically used to regulate the light intensity from the built-in lamps. But closing this iris diaphragm more than one-third deteriorates resolution. Using neutral density filters, which can be inserted between the light source and the objective or the substage condenser, is a better alternative.

If the microscope contains a condenser lens, it will be located directly beneath the stage and between the light and the specimen. However, for routine work, the condenser should not be used to adjust illumination, but should be set at or near its uppermost position and the light intensity adjusted as previously described.

Images are focused by moving the lens system (the body tube) in reference to the specimen (in most models the lens system is fixed and the stage moves instead). Coarse and fine adjustment knobs are used to accomplish this. In many microscopes, a black line or pointer seems to cut across the field of view. This pointer is typically made of hair or other fine material and enables a particular object or region of the material to be singled out during observation. A variety of pointers or counting grids may be added to the ocular lens.

OPTICAL PRINCIPLES. In light microscopy, the object observed is seen by the light it reflects or transmits. One of the primary differences between the low-magnification stereoscopic microscope and the high-magnification compound microscope is that the stereoscopic microscope typically uses reflected light and the compound microscope typically uses transmitted light.

Resolution. Although the microscope is a magnifying instrument, magnification alone does not necessarily improve the ability to observe details. The ability of a microscope to allow small objects to be distinguished is more important than its overall magnifying capabilities. The resolving power or resolution of an optical system is its ability to make clear and distinct the separate parts of an object. In other words, it is a measure of how clearly details can be seen with the microscope.

The resolving power of a lens system depends on the angle of the light cone gathered by the objective and the wavelength of light used. The unaided human eye has a resolving power of approximately 0.1 mm, which is determined by the limited number of light-sensitive cells comprising the retina of the eye. When images from two separate objects fall on the same light-sensing cell, the two images are perceived as one. A microscope coupled with the human eye, by magnifying an object and increasing the angle through which light is gathered, produces a lens system with a greatly improved resolving power.

However, the sensitivity of the human eye still places a limit on the resolving power of the eye–microscope system. The theoretical limit of resolution of the light microscope is approximately one-half of the wavelength of the light used. Because the human eye cannot perceive wavelengths shorter than approximately 400 nm (0.0004 mm, or 4 microns), the resolving power of a compound light microscope (with a 10X ocular and 100X objective) with the human eye as a receptor is approximately 200 nm. Use of immersion oil, with a 100X immersion objective fitted with an oiled condenser top, is needed to achieve this 200-nm resolution.

Magnification. Magnification is a measure of how big an object looks to the eye where life-size images are specified as 1. Magnification is typically written by a number followed by "X," which stands for times life size (for example, 10X means 10 times life size). The ocular lens on the microscope typically provides 10X magnification of the image projected by the objective lens. Therefore, total magnification is the magnification of the ocular times the magnification of the objective. For example, a 10X ocular lens in combination with a 40X objective would provide a total magnification of 400X.

Depth of Field. When using a microscope, the vertical distance that remains in focus at any particular setting is called the depth of field. This is a constant value for a given objective. If a specimen is thicker than this value, focusing up and down is required to clearly see the area being observed. This technique, called optical sectioning, can give valuable information on the three-dimensional shape of a specimen. When used with the fine-line adjustment knobs, the thickness of an object can be estimated. On many microscopes, the fine adjustment knobs are calibrated in measuring units of micrometers (0.001 mm, or 1 μm = micron).

OPERATING PROCEDURES. The following procedures are general operating guidelines for compound microscopes. The instruction manual for a particular microscope should always be consulted for specific procedures.

Step 1. Clean the ocular and objective, being sure to only use lens paper. Other cleaning materials may scratch the lens and leave fibers behind. The substage mirror can be cleaned with any clean paper tissue or cleared of dust with a dry air blower.

Step 2. Rotate the nosepiece so that the lower magnification (4 or 10X) objective is in place.

Step 3. If the microscope has a substage condenser, set it to its uppermost position.

Step 4. Open the condenser iris diaphragm fully (or place the largest hole of the disc diaphragm under the stage hole) and turn on the microscope lamp. If the microscope lamp has a rheostat, use a low rheostat setting when turning the light on or off. If the microscope is equipped with a substage mirror and an auxiliary lamp, adjust the positions of the lamp and mirror (use the flat mirror if one is present) until an evenly lit, circular field is obtained.

Step 5. If the microscope has a rotatable disc diaphragm, rotate the disc so that a middle size aperture is positioned in the light path. If the microscope has an iris diaphragm, remove the eyepiece, adjust the diaphragm opening three-quarters to four-fifths the size of the field seen through the observation tube, and replace the eyepiece.

Step 6. If a movable body tube is used to focus, while looking from the side, use the coarse adjustment knob to lower the lens system as far as it will go without touching the slide. Look into the ocular and focus slowly upward using the coarse focus knob. Once the specimen comes into view (if necessary, adjust the position of the slide), complete focusing with the fine adjustment knob.

If a movable stage is used to focus, while looking from the side, raise the stage by turning the coarse adjustment knob until the slide is approximately 3 mm from the objective. While looking through the ocular, slowly lower the stage by turning the coarse adjustment knob until the specimen comes into focus. Obtain a sharp focus by using the fine adjustment knob.

Step 7. To give a comfortable view of the specimen, adjust the light by adjusting the rheostat or by changing the distance between the lamp and the substage mirror.

Step 8. Note that objects will seem to be inverted (upside down and backwards). Move the slide from right to left and note which way the image appears to move. These inversions are caused by the optics of the microscope. Some practice will be required to become accustomed to these optical inversions.

Step 9. To view more detail, use the high-magnification objective, but always locate the specimen using the low magnification objective first. After centering and focusing the specific part of the specimen, switch to high magnification as described in Step 10.

Step 10. While watching from the side of the microscope, rotate the high magnification (40X) objective into position. It should not be necessary to adjust the distance between the stage and the body tube before making this change. Most microscopes have parfocal objectives; thus, if the image was in focus with the low-magnification objective, it should be nearly in focus under high magnification. Only a minor adjustment with the fine focus knob should be needed.

Step 11. Increase the light intensity on the specimen by either rotating the wider aperture of the disc diaphragm into position under the stage hold or increasing the aperature size of the condenser iris diaphragm. Note that at high magnification both the light level and the portion of the specimen visible within the microscope's field are less than at low magnification.

Operating Tips. Some laboratory analyses may require using a microscope for an extended period of time. To avoid fatigue, both eyes should be kept open when viewing the specimen. One exercise that can be used to master this technique is as follows:

While peering into the eyepiece, imagine staring at an image some distance below the level of the table. Then, with both eyes open, concentrate on the image with the eye that is at the eyepiece. The sooner this technique is learned, the more likely a headache will be avoided.

The following operating hints will help to ensure proper use of the microscope:

- Always check the cleanliness of the lenses before using the microscope. The extra time needed to do this will be worth the effort. Lens paper is the only material that should be used for this lens cleaning. For particularly dirty lenses or lenses caked with immersion oil, a few drops of ethanol or xylene on the lens paper followed by gentle blotting with a clean lens paper is recommended.
- To avoid ramming the specimen with the objective lens, damaging both the specimen and the lens, never focus downward while looking through the ocular. Always watch from the side while making significant adjustments. The best procedure is to lower the body tube (or raise the stage) until the objective almost touches the specimen. Then look into the ocular and slowly raise the body tube (or lower the stage) with the coarse focus knob until the image is visible. Complete focusing with the fine focus.
- Always watch from the side when changing objectives (especially when changing from 10X low magnification to 40X high magnification). If there is insufficient clearance between the objective and the specimen, do not force the objective into position. This would break the slide and possibly damage the objective.
- Develop a habit of continually varying light levels to find the level necessary in each case. Inexperienced microscopists tend to illuminate the specimen with excessive light. This is particularly unfortunate with compound microscopes because excessive light eliminates what little color or contrast the specimens possess.
- Be sure to keep the stage and viewing plates of the microscope clean and dry at all times. When finished, switch to the low-magnification objective, rack down or lower the mechanical stage fully, replace the dust cover, and return the instrument to a suitable storage space.
- When moving the microscope from one place to another, always hold it in an upright position. Also, use both hands, one under the base and the other to grip the body arm.

MEASURING OBJECTS. Frequently, the dimensions of the object being examined under the microscope are needed. The following two methods can be used to obtain estimates of the size of microscopic objects.

While either method can be used, method B is more accurate and recommended even though it requires special equipment.

Method A. This method approximates the size of an object by comparing it with the diameter of the field of view. To do this, the size of the field with the lower power objective in use must first be measured.

1. Place a short, clear plastic ruler over the opening in the center of the stage so that the scale is visible through the microscope.
2. Line up one of the vertical lines of the ruler so that it is visible at the left of the circular field of view.
3. Count the number of millimeters included from one side of the diameter of the field to the opposite side. This represents the size of the field of view for this objective. For example, if an object's diameter is one-third that of the field and the diameter of the field is 6 mm, then the diameter of the object is approximately 2 mm.
4. Turn the next higher magnification objective into place. Instead of measuring this field directly it will be more accurate to obtain the diameter as follows: divide the magnifying power of the higher power objective by the magnifying power of the lower power objective, then divide the diameter of the lower power objective field obtained previously by this quotient. For example, if the diameter with the 4X objective is 6 mm and for the 10X and the 40X lens, the field of view is calculated as follows:

$$10 \div 4 = 2.5 \quad 6 \text{ mm} \div 2.5 = 2.4 \text{ mm}$$
$$40 \div 4 = 10 \quad 6 \text{ mm} \div 10 = 0.6 \text{ mm}$$

Method B. This method accurately measures objects using an ocular micrometer, a small glass disc on which uniformly spaced lines are etched. The disc is inserted on the fixed diaphragm in the eyepiece of the microscope (Figure 3.2a). When an object is observed, the micrometer acts as a ruler whose image is superimposed on the object. However, because objectives vary in their actual magnification, the lines of the micrometer must first be calibrated for each objective of the specific microscope. This calibration is done with a stage micrometer, a special glass slide with uniform lines etched at known standard intervals. The usual distance between each large line is 0.1 mm and each small line is 0.01 mm (10 microns). These values will be labeled on the slide. If the stage micrometer is observed without the ocular micrometer in place, it will appear as shown in Figure 3.2b. When the ocular micrometer is in place, it will appear as shown in Figure 3.2c. To calibrate the ocular micrometer

1. Turn the ocular in its tube until the lines of the ocular micrometer are parallel with those of the stage micrometer. Match the lines at the left edge of the two micrometers by moving the stage micrometer (Figure 3.2d).
2. Calculate the actual distance between the lines of the ocular micrometer by observing how many spaces of the stage micrometer are included within a given number of spaces on the ocular micrometer; use the following simplified formula:

10 spaces on the ocular microscope = x spaces on the stage micrometer.

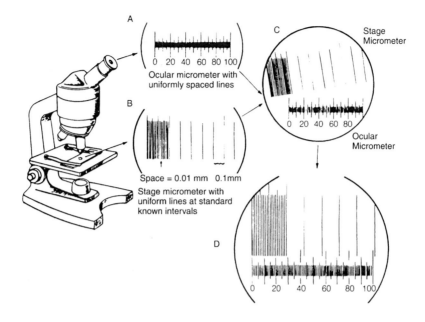

Figure 3.2   Using an ocular micrometer to determine the size of microscope objects (Method B).

Because the smallest spaces on the stage micrometer are known (printed on the slide) to be 0.01 mm apart, then

$$\text{Size of one ocular micrometer space} = \frac{(\text{x spaces})(0.01\text{mm/space})}{10 \text{ ocular micrometer spaces}}$$

3. Calibrate each objective lens in this way. Once the objective lenses are calibrated, all that is needed to determine the size of an object is the ocular micrometer. When viewing an object, simply count the number of ocular micrometer spaces it covers and multiply that number by the end size of the micrometer space calculated for the particular objectives used.

COUNTING CHAMBERS. In addition to measuring objects under the microscope, periodically the items under view must also be counted. A number of commercially available counting devices exist, including griddling membrane filters, the Sedgwick-Rafter counting cell, the Palmer-Maloney counting cell, the McMaster counting cell, and the hemacytometer (a counting chamber routinely used to make blood cell counts). Descriptions of most of these counting devices can be found in *Standard Methods* (APHA et al., 1989). Appendix A describes a modified application of the hemacytometer for use in enumerating filaments (Chapter 8) and parasitic ova (Chapter 11).

STEREOSCOPIC BINOCULAR MICROSCOPE

DESCRIPTION. The low-magnification stereoscopic binocular (or dissect-ing) microscope allows the observation of objects that are too large or too thick to be seen with higher magnifications but too small to be seen with the unaided eye. It also provides an opportunity to observe these objects in three dimensions. This type of microscope typically consists of two compound-lens systems that are focused simultaneously on the same region of the object; each eye views the field from a different angle. This arrangement provides the observer with binocular vision and increased depth perception with long working distance.

Most stereoscopic microscopes feature variable or zoom magnification. This is accomplished by varying the distance between the objective elements and ocular lenses. As with the compound microscope, total magnification is calculated by multiplying the separate magnification of the objective and ocular lenses.

In some models, magnification (zooming) is varied by rotating a knob of the microscope or by rotating a ring around the center of the body tube. Total magnification typically ranges from 4 to 60X and most microscopes are equipped with 10X ocular lenses.

Depending on the model, the microscope may have either an attached stage light with a separate transformer or a separate high-intensity lamp, a ring light, or a fiber optic illuminator. The lamp is typically positioned so that light reflects off the specimen and enters the lens system. The diffuse blue-light-emitting lamps used with compound microscopes typically are not appropriate for illuminating specimens on the stereoscopic microscopes. Reflected light is best for color and depth perception.

The optical systems of the stereoscopic microscope include erecting prisms, which eliminate the image inversion that occurs in the compound microscope. The relatively large depth of field, long working distance, and the erect image obtainable with the stereoscopic microscope make it particularly useful for manipulating objects while viewing them.

OPERATING PROCEDURES. General procedures for using the stereo-scopic binocular microscope are as follows. The instruction manual for the particular microscope used should always be consulted for specific proce-dures.

Step 1. Select the view plate that gives the greatest contrast between the specimen and the background. Usually the viewing plate on the microscope's stage has a black side and a white side.

Step 2. Place the specimen on the viewing plate, set the microscope at the lowest possible magnification, and correctly position the light source.

Step 3. Adjust the distance between the two oculars to match a comfortable viewing distance by moving the one adjustable ocular.

Step 4. Focus the adjustable ocular to compensate for eye vision differences.

Using the highest 200 M setting (microscopes may vary; check manufacturer's instructions) with one eye closed, look through the nonadjustable ocular and clearly focus the image using the focus knob. Then, using the lowest 200 M setting look through the adjustable ocular with the other eye closed and rotate the ocular's milled cuff to obtain a clear focus without using the focusing knobs.

SELECTING A MICROSCOPE

The compound and stereoscopic binocular microscopes described previously are considered standard equipment by many laboratories. In addition to these two types of microscopes, there are many accessories, modifications, or special microscopes available for particular applications (for example, to yield better contrast or resolution). Several considerations for selecting a microscope (or accessories) include

- The specimen (size range, live or dead, and stained or unstained) that will be examined.
- The type of illumination (transmitted or reflected, filtered or unfiltered, and special contrast techniques) that will be necessary.
- The magnification and resolution requirements of the objective to show the smallest detail of interest.
- Photographic requirements that are necessary.
- Budget constraints.

Today's microscopes and their accessory systems offer high precision and discrimination for a variety of laboratory demands. Several typically used microscopes are described in the following paragraphs, and Table 3.1 compares their operation, advantages, and disadvantages.

DARK-FIELD MICROSCOPE. The dark-field microscope shows border effects in transparent specimens and greatly improves their contrast. A special metal dark-field stop is inserted below the iris diaphragm of the microscope (a special dark field condenser is used). Only a hollow cone of light passes through the outer part of the condenser lens and illumination of the object is by reflected light, leaving the surrounding field dark. Loss of internal cell details occurs, however.

PHASE-CONTRAST MICROSCOPE. Phase contrast is of value in the study of transparent specimens. Structures in living cells may be seen because

Table 3.1 Comparison of different types of microscopes typically used.

Characteristics	Compound bright-field microscope	Dark-field microscope	Phase-contrast microscope	Fluorescence microscope	Electron microscope
Basic Principal	Transmitted light	Transmitted light	Transmitted light	Reflected UV, blue, or green light for excitation; light emitted in form of visible fluorescence	Beam of electrons
	Optical lenses	Optical lenses	Optical lenses and diffraction plates		Magnetic field lens
	Darker object on a bright background	Bright object on a dark background	Light waves in phase	Fluorescent object against a dark background	Photographic film or fluorescent screen needed to record image
Highest magnification	Visual 1000X	Visual 1000X	Visual 1000X	Visual 1000X	Up o 250 000X
Advantages	Cells can be viewed either alive or dead; simple operation	Cells can be viewed either alive or dead; no staining is necessary	Cells can be viewed either alive or dead; no staining is necessary	Improved specimen visibility; specimens can be rendered selectively visible by stains	Extremely high resolving power
	Low cost	High contrast	Moderately high contrast		
Disadvantages	Staining is usually required	Loss of internal details; need strong light	May introduce halos on the image, boundaries, and edges	Typically not used to observe living cells	Cannot observe living cells; required skill; expensive auxiliary equipment
				High cost	High cost

greater contrast is produced between them and their surroundings. This is accomplished by equipping an ordinary compound microscope with phase-contrast objectives and a special annular opening in the condenser and a transparent disc (called a phase plate) in the objective, thereby modifying a portion of the light that passes through the microscope. In doing so, the refractive index of the specimen is exaggerated and it becomes possible to distinguish structural details that vary only slightly in thickness or refractive properties. The phase-contrast microscope is used routinely for examining such organisms as filamentous algae, parasite eggs and cysts, and the filamentous bacteria that cause bulking.

INTERFERENCE-CONTRAST MICROSCOPE. The interference-contrast microscope provides another means of studying transparent objects and is based on principles similar to those of the phase-contrast microscope, but has the advantage of providing quantitative data. Interference microscopy permits detection of small, sharp changes in refractive index. The interference microscope combines a double-beam interferometer with a polarizing microscope. The polarization and wave interference produce greater contrast than is available with phase contrast and produce greater images in optical coloring and pseudo-three-dimensionality.

A special variation of the interference microscope is the Nomarski interference-contrast microscope, in which a single light beam passes through the object and the objective and is then divided into two interfering beams via a special birefringent prism. This microscope also includes polarizer and analyzer filters and a compensating prism in the substage condenser. The image obtained gives a characteristic relief effect and offers many advantages over ordinary phase-contrast optics, especially for thicker specimens. However, it is not as good as phase-contrast microscopy for observing filamentous microorganisms.

INVERTED MICROSCOPE. The inverted microscope, with the objective lenses below the stage, is used for examining materials requiring special handling. It allows investigators to see through the bottoms of petri dishes or flasks, typically with phase contrast.

POLARIZATION MICROSCOPE. The polarization microscope allows variation in the darkness and brightness and optical color of some otherwise invisible objects to be obtained. The optical principles of the microscope are based on the behavior of certain components of cells and tissues when they are observed with polarized light. The polarizing microscope differs from an ordinary light microscope in that two polarizing devices have been added: the polarizer and the analyzer. Both of these devices can be made from a sheet of polaroid film or, in former years, with Nicol prisms of calcite. The polarizer is mounted below the substage condenser and the analyzer is placed above the objective lens.

FLUORESCENCE MICROSCOPY. In fluorescence microscopy cells or tissues are examined under UV, blue, or green light and the components are identified by the fluorescence they emit in the visible spectrum.

Two types of fluorescence may be observed: natural fluorescence (autofluorescence), which is produced by substances typically present in the tissue, and secondary fluorescence, which is induced by staining with fluorescent dyes called fluorochromes that bind to cell structures or chemical components.

ELECTRON MICROSCOPE. The electron microscope uses a beam of electrons (with a wavelength greater than 0.5 mm). The object studied is flooded with electrons rather than light waves (Grimestone, 1977). Whereas ordinary light microscopes described above can magnify structures approximately 1000 times, the electron microscope can magnify them 250 000 times or more.

Transmission Electron Microscope. In the transmission electron microscope, a beam of electrons passes through the specimen and strikes a photographic or fluorescent screen. The specimen, embedded in plastic, must be cut in ultrathin sections so that the beam of electrons can pass through it. This type of electron microscope has been most valuable for studying the details of internal cell structure.

Scanning Electron Microscope. In the scanning electron microscope, the electron beam does not pass through the specimen, but causes secondary electrons to be emitted from the surface of the specimen, which has been coated with a thin film of metal. A recording of the emission from the specimen provides a three-dimensional picture.

*R*EFERENCES

American Public Health Association; American Water Works Association; and Water Pollution Control Federation (1989) *Standard Methods for the Examination of Water and Wastewater*. 17th Ed., Washington, D.C.

Gerardi, M.H., et al. (1983) Identification and Enumeration of Filamentous Microorganisms Responsible for Bulking Sludge. *Water Pollut. Control Assoc. Pa. Mag.*, **16**, 32.

Glase, J.C., et al. (1979) *Investigative Biology—A Laboratory Test*. Morton Publishing Co., Denver, Colo.

Grimestone, A.V. (1977) *The Electron Microscope in Biology*. 2nd Ed., Edward Arnold, Ltd., London.

Suggested Readings

Bradbury, S. (1976) *The Optical Microscope in Biology.* Edward Arnold, Ltd., London.

Needham, G.H. (1958) *Practical Use of the Microscope.* C.C. Thomas, Springfield, Ill.

Post, N., and Brenner, M. (1979) *Considerations in the Selection of a Microscope.* Am. Lab., **11**, 81.

van Norman, R.W. (1971) Microscopy. In *Experimental Biology.* 2nd Ed., Prentice Hall, Englewood Cliffs, N.J., 98.

Chapter 4
Bacteria

INTRODUCTION

WHAT ARE THE BACTERIA? Bacteria are the largest group of living organisms, in terms of biomass and, perhaps, numbers of species. They are the recyclers of natural and human-managed systems. As such, they mediate the biological processes that are used in wastewater treatment. The purpose of this chapter is to provide an overview of the bacteria.

Bacteria are single-celled organisms that belong to the kingdom Monera. Bacteria are prokaryotes as opposed to eukaryotes, which include fungi, protozoa, plants, and animals. Viruses are neither prokaryotes nor eukaryotes.

The prokaryotes differ from eukaryotes by the following primary characteristics:

- Prokaryotes are typically less complex.
- The DNA (genetic material, i.e., genes) of prokaryotes is not separated from the rest of the cell by a nuclear membrane.

- Prokaryotes do not have differentiated, membrane-bound organelles performing specialized cellular functions (e.g., chloroplasts and mito-chondria).
- Prokaryotes divide by binary fission, although many protozoa also reproduce this way.

Bacteria are smaller than eukaryotic cells, ranging in average size from 0.2 to 0.3 µm to 1 to 2 µm. Exceptions are the filamentous bacteria and cyanobacteria that can be larger than 100 µm and 5 to 50 µm, respectively. Bacterial growth rate is greater than that for eukaryotes. This is because smaller cells have a greater surface-to-volume ratio than larger cells. The larger membrane surface area allows a greater transport of nutrients and wastes into and out of the cell.

ROLE IN TREATMENT. Bacteria are responsible for most of the recycling that occurs in ecosystems. They have been termed the decomposers of the ecosystem. In this role, they grow (produce bacterial biomass) by breaking down (metabolizing) organic materials including feces; urine; and animal, plant, and bacterial remains into inorganic compounds such as nitrogen gas, nitrates, sulfates, phosphates, and carbon dioxide. Thus, bacteria recycle bound organic materials and energy by metabolizing them into inorganic compounds that can then be reused by other organisms for their growth and development.

Wastewater treatment facilities are actually human-engineered and human-managed ecosystems. Biological treatment processes rely on the natural recycling activities of bacteria found in natural ecosystems. In effect, waste-water treatment facilities condense and accelerate the bacteria-mediated recycling processes that occur naturally in ecosystems. Bacteria comprise approximately 90 to 95% of activated-sludge biomass.

CELL SHAPES, TYPES, AND SIZES. Bacterial cells occur in a variety of shapes, types, and sizes. There are three basic shapes and seven unusual types. Sizes range from 0.2 to 0.3 µm to greater than 100 µm. The basic shapes are coccoid, bacilloid, and spirillar (Figure 4.1). The cocci are spheri-cal or round, such as *Streptococcus* spp. The bacilli are rod shaped, such as

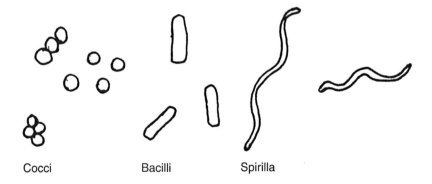

Cocci   Bacilli   Spirilla

Figure 4.1 Common bacteria shapes.

Bacillus subtilus. The spirilla are spiral shaped, for example, *Vibrio cholera* and *Spirillum volutans.*

The unusual types of bacteria are characterized by uncommon shapes, arrangements, and movements (Figure 4.2). Sheathed bacteria are filamentous

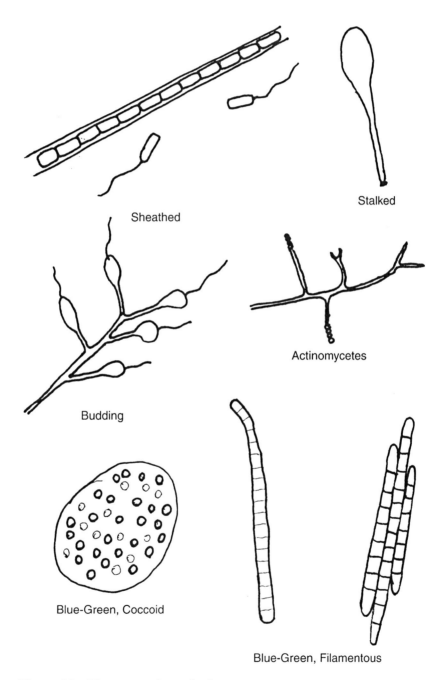

Figure 4.2   Uncommon bacteria shapes.

and surrounded by a tube-like structure called a sheath. Rod-shaped cells inside the sheath can leave it becoming flagellated (developing a tail-like extension) and free swimming. These bacteria are typical in polluted environments and wastewater. *Sphaerotilus*, *Leptothrix*, and *Crenothrix* are examples of sheathed bacteria.

Stalked bacteria are club-shaped and flagellated. The slender extension or stalk ends in a holdfast structure that allows the cell to attach to surfaces. Rosettes may form when cells attach to each other. *Caulobacter* is typical in low-organic-content aquatic environments. *Gallionella*, which forms a twisted stalk or ribbon, is found in iron-rich waters and the metal pipes of water distribution systems.

Budding bacteria attach to a surface, form hyphae (filaments), and multiply by budding off cells from the ends of the hyphae. Buds develop a flagellum, detach, and swim (swarm) to a new surface where they settle and repeat the process. *Hyphomicrobium*, found in soils and water, is an example of this type.

Gliding bacteria are filamentous and exhibit slow, smooth movement (gliding) over solid surfaces. Examples include *Beggiatoa* and *Thiothrix*. *Thiothrix* can contain sulfur granules giving it a striking string-of-(golden) pearls appearance under the microscope. It can also form rosettes.

Bdellovibrio are small (0.2 to 0.3 µm), flagellated bacteria that parasitize larger, Gram-negative bacteria. They can form plaques (clear areas) on a lawn of host bacteria grown on plated media.

Actinomycetes are filamentous bacteria similar in appearance to fungi. They are characterized by mycelial growth; that is, growth by branching filaments. They are within the bacterial size range with a filament diameter of approximately 0.1 µm. Most produce spores, some cause a characteristic earthy odor, still others produce antibiotics effective against other bacteria. Examples include *Streptomyces*, *Micromonospora*, and *Nocardia*. *Nocardia* is a primary cause of activated-sludge foaming.

Cyanobacteria, often called blue-green algae, are bacteria that appear blue-green and can photosynthesize because of the presence of phycocyanin, phycoerythrin, and chlorophyll a. They occur as single (unicellular), colonial (multiple cells), or filamentous organisms. They are quite tolerant of extreme conditions (e.g., pollution) and are typically found in polluted systems and algal blooms.

*P*ARTS *OF THE CELL*

The various parts of the cell are shown in Figure 4.3.

CELL WALL. All bacteria, except mycoplasma, are surrounded by a cell wall. It gives rigidity to the cell, maintains cell shape, and resists excess osmotic (fluid) pressures. The cell wall consists primarily of a mucopolysaccharide; that is, peptidoglycan. Peptidoglycan is a long chain (polymer) of

---

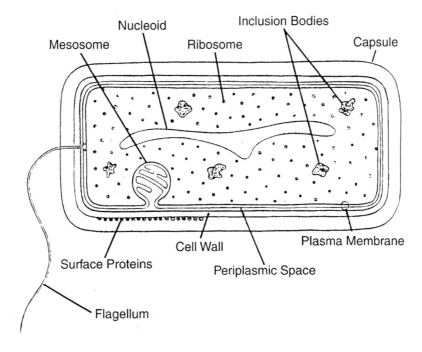

Nucleoid

Inclusion Bodies

Mesosome   Ribosome   Capsule

Cell Wall   Plasma Membrane

Surface Proteins   Periplasmic Space

Flagellum

Figure 4.3   Morphology of a Gram-positive bacterial cell. The majority of the structures shown here are found in all Gram-positive cells. Only a small stretch of surface proteins has been included to simplify the drawing; when present, these proteins cover the surface. Gram-negative bacteria are similar in morphology (Prescott et al., 1990).

amino sugars. The Gram stain is specific for cell walls and differentiates bacteria as Gram positive or Gram negative based on cell wall staining characteristics. In Gram-positive bacteria, peptidoglycan comprises as much as 50% or more of the wall's dry weight. By contrast, in the Gram-negative bacteria, peptidoglycan is typically less than 10%, sometimes as little as 1%. Instead, Gram-negative bacteria contain lipoproteins and lipopolysaccharides; that is, fats bonded to proteins and sugars. During Gram staining, the alcohol rinse (decolorizing step) dissolves the lipoproteins and washes out the primary stain (the Gram-positive reaction). The counter stain (the Gram-negative reaction) can then stain the cell.

CYTOPLASMIC (PLASMA) MEMBRANE. Within the cell wall, the bacterial cell is surrounded (delimited) by a cytoplasmic membrane (also called the plasma or cell membrane). Structurally, it consists of a phospholipid bilayer with embedded proteins (Figure 4.4). Phospholipids are molecules that contain a phosphoric acid head and fatty acid tails. The phosphoric acid component makes the phospholipid soluble in water (hydrophilic). The fatty acid tails are fat soluble (hydrophobic). The heads project to the outside

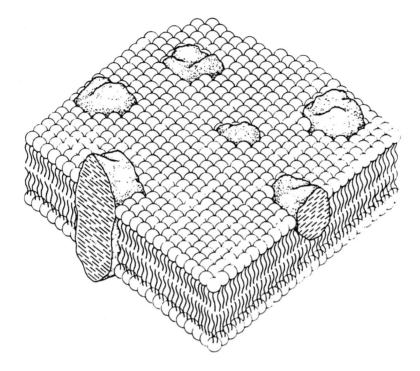

Figure 4.4 Diagram of the fluid mosaic model of membrane structure showing the integral proteins floating in a lipid bilayer. Some of these integral proteins extend all the way through the lipid layer. The hydrophilic ends of the membrane lipids are represented by small spheres. See text for further explanation (Prescott et al., 1990).

and inside of the bacterial cell and the tails are between the heads, within the cell membrane. The embedded proteins act, in part, as channels for the passage of certain substances. The membrane is selectively permeable; that is, permitting the passage of certain substances such as proteins, sugars, wastes, and liquids. Substances cross the membrane either by diffusion or active transport. Diffusion does not require energy use by the cell (hence the term passive diffusion), and is driven by concentration differences across the membrane. In active transport, the cell uses energy to transport substances across the membrane against the concentration gradient (hence the term active transport).

GLYCOCALYX. Some bacteria produce a capsule, called a glycocalyx, of extracellular polymeric substances (primarily polysaccharides) outside the cell membrane and cell wall. Capsules are important because they (a) affect pathogenic virulence (i.e., the ability to cause disease); (b) aid in adsorption (attachment) to surfaces, as in the case of pipe fouling; (c) help protect the

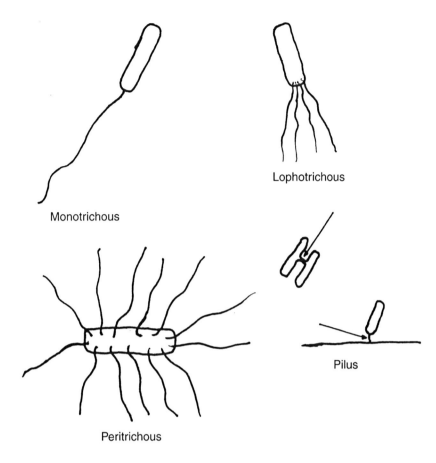

Figure 4.5   Flagella and pili.

cell from predation; (d) help the cell resist desiccation (dehydration); and (e) aid in microbial flocculation, such as in the activated-sludge process.

FLAGELLA AND PILI. Motile bacteria typically exhibit flagella, long hairlike structures that are used for propulsion (Figure 4.5). Flagella may be found as a single, polar (at one end of the cell) flagellum as in the monotrichous condition of *Vibrio comma*. A bundle of flagella may be located at one end of the cell, called the lophotrichous condition, as in *Spirillum volutans*. Alternatively, flagella may be numerous and distributed around the cell, as in the peritrichous condition of *Escherichia coli*.

Pili (pilus, singular) are short, thin structures that are used by cells to attach to surfaces. A specialized pilus is used during conjugation (sexual reproduction in bacteria). Genetic material is transferred through the pilus between the conjugating cells.

STORAGE PRODUCTS. Some cells store chemical compounds internally. These stored materials, called inclusions, may serve as energy sources or as cell-building materials. Primary inclusion products include sulfur granules in filamentous sulfur bacteria (e.g., *Beggiotoa* and *Thiothrix*); poly-β-hydroxy-butyric acid (PHB, found exclusively in prokaryotes), which stains with Sudan Black; and polyphosphate (poly P) granules (e.g.,volutin granules), which stain with Neisser stain and are found in poly P bacteria such as *E. coli*, *Pseudomonas*, and *Beggiotoa*.

GAS VACUOLES. Vacuoles are membrane-bound spaces within cells that contain gas. They are found in cyanobacteria, photosynthetic bacteria, and halobacteria (salt-tolerant bacteria). Gas vacuoles provide buoyancy to regulate the cell's location in the water column. For example, in highly turbid water, cyanobacteria float to the surface because of their gas-filled vacuoles, allowing them maximum use of available sunlight.

ENDOSPORES. Endospores are a resistant stage in the life cycle of certain bacteria; for example, *B. subtilis* and *Clostridium perfrigens*, and are formed when environmental conditions are unfavorable for the normal, vegetative state. Spores are formed within the cell and are very resistant to heat, desiccation, radiation, and chemicals. Endospores are resistant to the standard disinfection processes used in drinking water and wastewater treatment facilities and require more advanced disinfection methods for inactivation. Favorable physical and chemical conditions trigger spore germination and the return to the vegetative state.

DNA. Bacteria do not have a membrane-bound nucleus that contains their DNA as do eukaryotes. The DNA exists as a circular molecule contained within the cell. Smaller, circular DNA molecules, called plasmids, may also occur. Plasmids are important because they may contain genes that make the bacterium resistant to certain antibiotics. During conjugation (union), bacteria can share these antibiotic-resistant genes with each other. This explains why there are now antibiotic-resistant strains of bacteria, such as *Streptococcus* spp., *Mycobacterium tuberculae*.

CLASSIFICATION OF BACTERIA

In addition to classifying bacteria based on their evolutionary relationships, morphology, and biochemical characteristics (traditional taxonomy), bacteria may be classified based on their source of energy, oxygen requirements, and temperature preferences. These latter three classifications are not mutually exclusive.

AUTOTROPHIC AND HETEROTROPHIC BACTERIA. Autotrophs (or self feeders) use carbon dioxide as their source of carbon for growth. Photoautotrophs, such as cyanobacteria, obtain energy from sunlight and carbon from carbon dioxide. Chemoautotrophs obtain energy from inorganic compounds, such as ammonia and hydrogen sulfide, and carbon from carbon dioxide. Chemolithotrophs are bacteria that obtain energy by oxidizing inorganic nitrogen or sulfur compounds, or methane or methanol. Most chemoautotrophs are chemolithotrophs and are important in nutrient recycling and wastewater treatment. For example, *Nitrosomonas* and *Nitrobacter* are examples of the chemolithotrophs that effect nitrification of wastewater.

Heterotrophs (or other feeders) obtain energy from organic compounds. Chemoheterotrophs obtain energy and carbon from organic compounds. Most chemoheterotrophs are also chemoorganotrophs that break down (metabolize) organic matter and are necessary for secondary biological treatment processes and both aerobic and anaerobic digestion processes.

AEROBIC, FACULTATIVE ANAEROBIC, AND ANAEROBIC BACTERIA. Bacteria are classified based on how they metabolize matter. The metabolism of matter releases electrons that are ultimately transferred to a final electron acceptor. Bacteria can use various substances as a final electron acceptor. Those that use free oxygen are called aerobes; that is, they undergo aerobic respiration. Those that can use combined-oxygen substances (such as nitrate, sulfate, and carbonate) when free oxygen is absent are called facultative anaerobes; that is, they undergo anaerobic respiration in the absence of oxygen. Bacteria that cannot use free or combined oxygen are called anaerobes and undergo anaerobic fermentation. In this case, organic compounds such as alcohols and organic acids are used as final electron acceptors.

Chemoheterotrophic bacteria that require oxygen to carry out their life processes are called aerobic or aerobes. Some are strict aerobes in that they have an absolute requirement for oxygen to survive, although some can survive at low oxygen concentrations. Most of the bacteria important for wastewater treatment are aerobes and facultative anaerobes.

Facultative anaerobes can survive and grow in the presence or absence of oxygen provided there is a suitable carbon source available. Many intestinal bacteria are facultative anaerobes. Anaerobes cannot use oxygen for their metabolic processes. Strict anaerobes cannot survive in the presence of oxygen because oxygen poisons their metabolism. Some aerotolerant anaerobes can survive but not grow in low concentrations of oxygen. Anaerobic bacteria play a significant role during anaerobic digestion of sludges.

PSYCHROPHILIC, MESOPHILIC, AND THERMOPHILIC BACTERIA. Bacteria are also classified by the temperature ranges that are optimal for their survival and growth. Psychrophilic bacteria grow optimally at temperatures from approximately 0 to 20 °C (32 to 68 °F). Mesophilic bacteria grow best at approximately 25 to 40 °C (77 to 104 °F). Mesophiles are important in mediating secondary biological treatment and in initiating composting (the thermogenic phase). Thermophilic bacteria require temperatures of 45 to 50 °C (113 to 122 °F) or more for optimal growth. Thermophiles are important in the breakdown of carbohydrates and proteins that occurs during the composting and sludge digestion processes.

ORIGIN OF WASTEWATER TREATMENT BACTERIA

The bacteria that mediate wastewater treatment occur naturally in aquatic systems (rivers, streams, and lakes) and soils. Originally, wastewater treatment occurred by natural means after sewers discharged municipal and industrial wastewater plus stormwater runoff into surface water bodies. In coastal areas, wastewater was mixed and diluted by tides and currents. Natural systems were able to break down and recycle (i.e., assimilate) these small discharges when the contributing human population was small. With the beginning of the industrial revolution (after the Civil War in the United States), urban populations increased tremendously. The resulting increase in the quantity and change in quality of wastewater overwhelmed these natural systems. Malodorous sludge deposits accumulated below sewer outfalls and urban waterways and streams became, in effect, open sewers. These conditions prompted public demand for wastewater treatment and better sanitary practices during the 1880s and 1890s.

Mechanical methods for wastewater treatment began at the beginning of the 20th Century (1900 to 1920). At first, treatment consisted of screening and settling of solids (i.e., primary treatment). By the second decade of the century, activated-sludge systems (biological or secondary treatment) were in operation in the United Kingdom. In the 1920s to 1930s, biological (secondary) treatment of the fixed-media type were being used in the United States. Biological treatment relied on naturally occurring bacteria found in streams and rivers that received and assimilated wastewater. Many investigations into the assimilative capacity of streams were conducted during this early period because streams were still relied on to assimilate or treat much of the raw wastewater and all of the primary treatment effluent being discharged. When fixed-media systems were added to the treatment process, they were started by seeding with a sample from a stream or river sludge deposit. The bacteria in the sample initiated and sustained the biological treatment of the wastewater. Typical treatment bacteria included *Zooglea, Pseudomonas, Comomonas, Flavobacterium, Alcaligenes, Bacillus, Achromobacter, Corynebacterium, Acinetobacter, Aeromonas,* the filamentous *Sphaerotilus, Beggiotoa, Thiothrix,* and the nitrifying *Nitrosomonas* and *Nitrobacter.* During the 1920s to the present, activated-sludge systems were developed to treat larger wastewater flows than could be effectively treated by the fixed-media processes. These systems also relied on the same bacteria as used in fixed-media treatment. Activated-sludge units were seeded from fixed-media systems. Today, new activated-sludge systems or those recovering from significant process upset may be started (or restarted) by seed from well-functioning activated-sludge systems or more slowly, without any seed.

GROWTH ENVIRONMENT AND DOMINANCE

The growth environment (conditions) determines which bacteria are active and dominant in a system at any given time. The prevailing conditions determine the type, extent, and speed of the biological processes that occur. Those bacteria that find the existing conditions most conducive to their survival and growth will be the ones that dominate those conditions. As conditions change, there is a succession of microbiological species and dominant types change and are succeeded by others better adapted to the changed conditions. This is the reason that intestinal bacteria are not the primary organisms in wastewater treatment systems.

Secondary treatment processes provide a variety of managed growth environments with characteristic types of dominating organisms. Typically, the secondary treatment growth environment consists of highly oxygenated fixed-media processes (trickling filters and rotating biological contactors), oxidation lagoons and ditches, and activated-sludge basins. In effect, natural recycling processes are accelerated in these human-engineered and human-managed ecosystems by providing bacteria with abundant nutrients (wastewater) and oxygen (pumped or assisted aeration; e.g., disk rotation or forced aeration of activated-sludge mixed liquor).

Biological treatment processes provide an oxygenated and nutrient-rich medium in which aerobic and facultative anaerobic bacteria dominate. Aerobes and facultative anaerobes metabolize the organic matter and convert it to bacterial biomass (cells). Bacteria clump together as floc particles suspended in the vigorously aerated mixed liquor of activated-sludge systems or attach to solid surfaces as biofilm in fixed-media systems. In the activated-sludge process, aerobes and facultative anaerobes dominate in smaller floc with a greater surface-area-to-volume ratio. Larger floc particles may contain internal anaerobic areas that contain methanogens and other anaerobes. Biofilm is relatively thin, allowing oxygenation by exposure to air. Thus biofilm is dominated by aerobic and facultative anaerobic bacteria.

Excess biomass from both processes is condensed and settled as sludge. Biofilm sludge and wasted (excess) activated sludge are sent to solids-handling processes (e.g., digestion, dewatering, drying, landfilling, composting, land application). The remaining activated sludge is returned to the head of the secondary treatment works to seed incoming primary effluent. Biofilm sludge is not recycled to the head of the process.

Waste stabilization lagoons are dominated by aerobic and facultative anaerobic bacteria. Bacterial sludge accumulates by natural settling. Periodic dredging is required to remove excess biomass, which is handled as described above.

Nitrification is the process in which nitrifying bacteria; that is, autotrophic chemolithotrophs, metabolize ammonia to nitrate. Because nitrifying bacteria grow slowly and are inhibited by high concentrations of nutrients (high biochemical oxygen demand), chemoheterotrophs are dominant under ordinary biological treatment conditions. However, *Nitrosomonas* and *Nitrobacter* can become dominant and nitrify a wastewater when the mean cell residence time is long, oxygen concentration is suitable, and there is sufficient wastewater treatment retention time.

Anoxic and anaerobic conditions are necessary for biological denitrification and the first step of phosphorus removal. Under these conditions, denitrifying bacteria such as *Pseudomonas, Alcaligenes, Bacillus, Spirillum,* and *Acinetobacter* are active in converting nitrate to nitrogen and nitrous oxide gases. During the first step of phosphorus removal, conditions favor *Pseudomonas, Moraxella, Acinetobacter, Aeromonas*, and other bacteria that will release inorganic phosphorus to the medium. The second step is aerobic. Aerobes and facultative anaerobes will store phosphorus, thus removing it from the wastewater.

REFERENCES

Prescott, L.M.; Harley, J.P.; Klein, D.A. (1990) *Microbiology*. Wm. C. Brown, Dubuque, Iowa.

SUGGESTED READINGS

Bitton, G. (1994) *Wastewater Microbiology*. Wiley-Liss, Inc., New York.

Hammer, M., and Hammer, M., Jr. (1996) *Water and Wastewater Technology*. 3rd Ed., Prentice-Hall, Inc., Englewood Cliffs, N.J.

Jenkins, D.; Richard, M.G.; and Daigger, G.T. (1993) *Manual on the Causes and Control of Activated Sludge Bulking and Foaming*. 2nd Ed., Lewis Publishers, Chelsea, Mich.

O'Leary, W., et al. (Eds.) (1989) *Practical Handbook of Microbiology*. 2nd Ed., CRC Press, Boca Raton, Fla.

Water Pollution Control Federation (1987) *Activated Sludge*. Manual of Practice No. OM-9, Alexandria, Va.

Water Pollution Control Federation (1989) *Activated Sludge Microbiology*. Alexandria, Va.

Water Environment Federation (1994) *Wastewater Biology: The Life Processes*. Special Publication, Alexandria, Va.

Chapter 5
Protozoa

INTRODUCTION

Protozoa, which means *first animals*, are microscopic, eukaryotic organisms, most of which are single-celled (though some are colonial) and heterotrophic (i.e., requiring organic molecules for energy and incapable of synthesizing their own food). They are a diverse assemblage grouped together in the kingdom Protozoa, separate from other kingdoms such as those containing plants, animals, and fungi.

Protozoa are tiny, typically measuring 5 to 1000 μm in size, and most are visible only with the aid of a microscope. There is considerable morphological and physiological diversity within the group. Because actively feeding protozoa need water, all free-living (nonparasitic) protozoa are essentially aquatic, living in freshwater (including soil), brackish, and marine environments.

Aerobic wastewater treatment plants (WWTPs) have been colonized successfully by protozoa, and their presence was noted almost as soon as each process was developed. In the past three decades the significance of these organisms has begun to emerge and it is now known that they play an essential role in biological treatment processes. Furthermore, they may also be used as reliable biological indicators of effluent quality and plant performance.

The processes in which protozoa are particularly abundant include trickling filters where there are a great variety of species; activated-sludge processes, where there can be as many as 50 000 organisms per milliliter, which is approximately 5% of the dry weight of suspended solids in the mixed liquor; rotating biological contactors (RBCs), where large numbers of ciliates and amoebae are found, mostly associated with the surface biofilm; oxidation lagoons, where considerably fewer protozoa are found than are algae; and constructed wetlands, although few data are currently available.

CHARACTERIZATION

TAXONOMIC IDENTIFICATION. There are three types of keys available for identifying protozoa in wastewater treatment processes: those originally written for the specialist in protozoan taxonomy (e.g., Lee et al., 1985); those written for the general biologist, undergraduate, and amateur microscopist (typically limited to identification of common genera; e.g., Finlay et al., 1988); and those written specifically for the identification of protozoa in treatment processes (e.g., Foissner and Berger, 1996, and Hänel, 1979).

CLASSIFICATION AND DESCRIPTION. Traditionally, protozoan classification has been based largely on cell morphology (or structure) and the means of locomotion (or movement), as revealed by the light microscope. Among the exclusively parasitic groups, spore structure is also an important feature. Based on such observations, four primary protozoan groups were historically recognized: flagellates, amoebae, ciliates, and sporozoa (obligate parasites that produce spores).

In recent years protozoan classification has been revolutionized as modern techniques such as electron microscopy and molecular analysis have been used. Consequently, of the four primary groups previously recognized, only the ciliates remain; the others have been divided and redistributed throughout (or in some cases removed from) the kingdom Protozoa. The situation remains confused and currently there is no generally accepted classification scheme for the protozoa, although Corliss (1994) offers a useful interim scheme. At the time of this writing, publication of the 2nd edition of the Society of Protozoologists' *Illustrated Guide to the Protozoa* is therefore eagerly awaited.

In the absence of a stable and generally accepted classification scheme, the protozoa of significance to wastewater treatment processes will here be treated as belonging to their traditional groups (flagellates, amoebae, and ciliates), although it should be kept in mind that the first two groups have no taxonomic significance.

Flagellates. The flagellates are characterized by the possession of one or more flagella (long, tapering, hair-like appendages), which act as organelles of locomotion and feeding. The flagella typically beat with an undulating or whiplash motion that propels the organism through the water and also creates feeding currents to draw food particles toward the organism. Typically, where two flagella are present, one may project forward and the other trail behind. Often, the organism's flagella are longer than its body. The flagellates are now distributed among at least four kingdoms, including the stramenopiles, protozoa, fungi, and plants.

Some protozoan flagellates, especially members of the euglenid and dinoflagellate groups, possess plastids containing chlorophyll and are wholly or partially photoautotrophic (photosynthetic; i.e., capable of absorbing sunlight and using carbon dioxide to synthesize food). Such flagellates form a significant part of the protozoan communities in oxidation lagoons and are also frequently encountered in the upper layers of trickling filters (Figure 5.1a).

Other flagellates are colorless and heterotrophic, obtaining their food either as particles or in a dissolved form. Colorless flagellates are found in all types of treatment processes, perhaps the most frequently encountered types being members of the Kinetoplastida (Figure 5.1b).

Amoebae. The primary characteristic of the amoebae is their possession of pseudopodia, retractile processes that serve as organelles of locomotion and feeding. There is a considerable diversity of structure in the amoebae, particularly in the character of any shell or skeletal material that may be present, and in the type of pseudopodium (for example, broadly lobed, needlelike, or reticulate). The best contemporary guide to the amoebae is that of Page (1988).

Amoebae found in wastewater treatment processes belong to one of two groups, Rhizopoda or Heliozoa, although the latter is relatively uncommon. The Rhizopoda may be naked (without a shell) or testate (with a shell). Figure 5.1c shows an example of a naked amoeba. The cell matrix (the cytoplasm) is divided into a clear outer layer (the ectoplasm) that surrounds a granular inner area (endoplasm). Within the endoplasm are the nucleus, contractile vacuole, food vacuoles, storage granules, and other inclusions. Other groups of

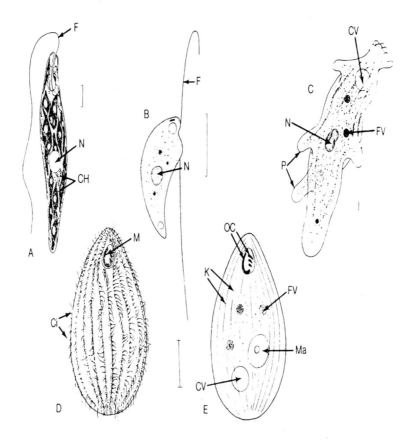

Figure 5.1   Representatives of the primary protozoan groups found in wastewater treatment processes (scale bars = 10 μm): (a) *Euglena gracilis*, **(b)** *Bodo caudatus*, **(c)** *Ameoba proteus*, **(d)** *Tetrahymena pyriformis* **ciliation, and (e)** *Tetrahymena pyriformis*. **Ci: cilia, Ch: chloroplasts, CV: contractile vacuole, F: flagellum, FV: food vacuole, K: kineties, M: mouth, Ma: macronucleus, P: psuedopodium, and OC: oral cilia.**

amoebae, notably the testate amoebae, possess shells that may be proteinaceous, agglutinate, siliceous, or calcareous in composition.

In wastewater treatment processes, naked and testate amoebae are fairly equally well represented. Little work has been carried out on the identification of the naked amoebae. They are normally present in low numbers but can, on occasion, become dominant. Although no more common than naked amoebae, testate amoebae are perhaps more easily noticed because of their distinctive test (shell), which can remain intact long after the amoeba has died.

Ciliates. The Ciliophora (ciliates) form a natural group distinguishable from other protozoa by a number of specialized features, including the possession of cilia (short hair-like processes) at some stage in their life cycle, the presence of two types of nuclei, and a unique form of sexual reproduction called conjugation.

Structures of representative ciliates are shown in Figures 5.1d and 5.1e. The body surface is covered with cilia, which are mostly aligned in rows called kineties. The pattern of kineties is interrupted in the region of the mouth where there may be specialized oral cilia used for feeding. There are two types of nuclei, a large conspicuous vegetative macronucleus and a small inconspicuous sexual micronucleus.

The primary ciliate groups found in wastewater treatment processes include gymnostomes (without prominent oral ciliation, Figure 5.2a), cyrtophorids and nassulids (with a cytopharyngeal basket; i.e., a specialized feeding structure that, in the case of nassulids, is for the ingestion of filamentous prey, Figure 5.2b), hymenostomes (with specialized oral ciliation, Figure 5.2c), hypotrichs (with compound ciliary organelles called cirri, Figures 5.2d, 5.2e, and 5.2f), and peritrichs (typically attached to a substratum by a stalk, Figure 5.3).

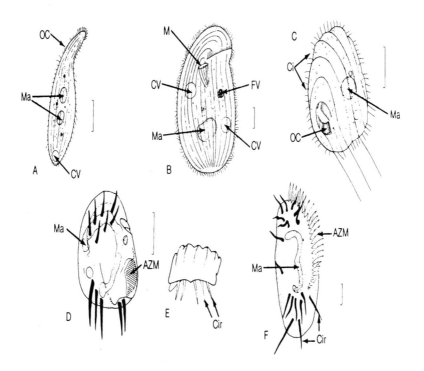

Figure 5.2 Common ciliates in aerobic treatment processes (scale bars = 10 μm): (a) *Trachelophylum pusillum*, **(b)** *Chilodonella uncinata*, **(c)** *Cinetochilum margaritaceum*, **(d)** *Aspidisca cicada* **ventral view, (e)** *Aspidisca cicada* **posterior view, and (f)** *Euplotes affinis*. **AZM: adoral zone of membranelles, Ci: cilia, Cir: cirri, CV: contractile vacuole, M: mouth, Ma: macronucleus, and OC: oral cilia.**

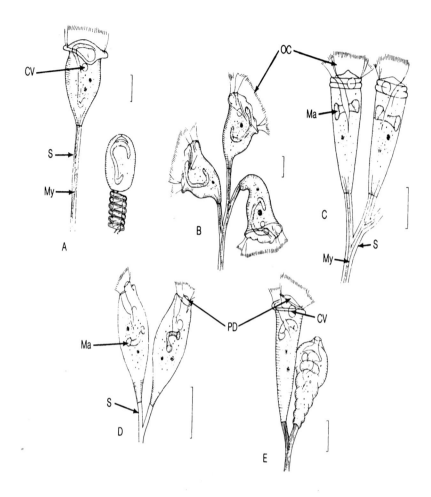

Figure 5.3 **Common peritrich ciliates in aerobic treatment processes (scale bars = 25 μm): (a)** *Vorticella convallaria*, **(b)** *Carchesium polypinum*, **(c)** *Zoothamnium* sp., **(d)** *Opercularia coarctata*, **and (e)** *Epistylis plicatilis*. **CV: contractile vacuole, Ma: macronucleus, My: myoneme, OC: oral cilia, PD: peristomial disc, and S: stalk.**

*F*EEDING PROCESSES

Protozoa found in WWTPs use one or more of three different feeding strategies: photoautotrophy, osmotrophy (ingestion of dissolved organic matter), and phagotrophy (ingestion of particulate food). The photo-autotrophs are all flagellates and are primary producers capable of absorbing sunlight and fixing carbon dioxide. The resultant organic compounds are stored and become available to predators, including other protozoa. In addition to light, they require a number of inorganic ions and trace elements, all of which are present in wastewater. Light is likely the most limiting factor for these organisms, consequently they will not typically be present in those treatment processes where light cannot penetrate.

Osmotrophic protozoa, primarily flagellates, use organic complexes dissolved in the wastewater. These protozoa may require vitamins, which many bacteria are able to produce. They compete with the heterotrophic bacteria for the organic complexes in wastewater.

The phagotrophs principally feed by ingesting particulate food matter, but their nutritional requirements are complex and have been completely determined in only a few species. Most protozoa in WWTPs are bacterivores, feeding either on dispersed or surface-associated bacteria. There are numerous reports that show that not all bacteria serve as a complete diet when supplied alone. It is also known that certain bacterial strains, particularly those that produce pigments, are toxic to amoebae and ciliates.

Other phagotrophic protozoa include saprotrophs, which ingest detrital matter, and carnivores, which feed on other protozoa. Wastewater treatment processes often contain carnivorous ciliated protozoa, which typically prey on attached peritrich ciliates. Suctorian ciliates are also found and these feed on a range of other protozoa. Apart from the true carnivorous ciliates, several can become cannibals.

REPRODUCTION AND GROWTH RATES

In free-living protozoa, the typical life cycle involves growth, or increase in cell size, followed by binary fission, which is a form of asexual reproduction. In binary fission, the cell mass is passed on to the succeeding generation as two new individuals and the biomass is retained within the population. Other forms of asexual reproduction include budding (unequal fission) and multiple fission (which gives rise to many daughter cells). Sexual reproduction also occurs in most groups. Typically, species that feed on bacteria show increasing growth and reproductive rates as food concentration increases. In activated-sludge processes, for example, ciliates can divide every 4 to 5 hours. Other groups such as the testate amoebae have much slower growth rates which are typically measured in days.

PROTOZOAN POPULATIONS IN WASTEWATER TREATMENT PROCESSES

More than 400 species of protozoa occur in wastewater treatment processes. Some of the more frequently encountered ones are listed in Table 5.1. The ciliates are almost always numerically the dominant protozoa and, as Table 5.1 indicates, are typically represented by the greatest variety of species. A total of approximately 250 species of ciliates have been reported in wastewater treatment processes, of which perhaps only one-third are typically found.

From Table 5.1 it can be seen that the protozoan fauna of trickling filters, activated-sludge processes, and RBCs have many species in common. There are more species simultaneously present in filters and RBCs and this is probably because of the greater variety of habitats that fixed-growth reactors offer. The fauna of oxidation lagoons are rather different and this probably reflects the difference between the former three totally aerobic systems and the latter, which is at least partially anaerobic. Few data are available for constructed wetlands, although aerobic and anaerobic species are both known to occur.

A survey of protozoa from activated-sludge processes and trickling filters in the United Kingdom was carried out in the 1970s to define which protozoa were most frequently found in these two aerobic processes (Curds, 1992). The results of this study, summarized in Table 5.2, indicate that all of the most important protozoa found in the two processes are ciliates. Most of them are sedentary, attaching themselves directly to the microbial films by means of a stalk. A few are crawling forms. All are known to feed on bacteria.

SPATIAL DISTRIBUTION. In fixed-growth processes such as trickling filters and RBCs, different organisms tend to predominate at different positions. This is because the wastewater is purified as it passes through the system, resulting in different conditions occurring at the beginning of the process compared to those at the end.

In trickling filters there is a restricted protozoan fauna at the surface, typically dominated by flagellates and amoebae, which changes to one of greater species diversity further down, where ciliates are more abundant. This vertical stratification of organisms in a filter bed depends on many factors, the two most important of which are thought to be food supply and environmental conditions. Basically, those organisms that can use soluble organic substrates are found near the surface, bacterivores predominate in the middle, and carnivores at the bottom. Similarly, there is a change from those that can tolerate organic-rich, low dissolved oxygen (DO) conditions near the surface to those that thrive in high DO and diminishing organic intensities towards the bottom.

Table 5.1 Some protozoa typically found in wastewater treatment processes (adapted from Curds, 1992).[a]

Protozoan species	Trickling filter	Activated sludge	RBC	Oxidation lagoon
Flagellates				
Bodo spp.	*	0	*	*
Pleuromonas jaculans	*	0	*	0
Euglena spp.	*	—	—	*
Peranema trichophorum	*	*	*	*
Choanoflagellates	—	—	—	—
Amoebae				
Amoeba proteus	0	0	*	*
Acanthamoeba castellanii	0	0	0	*
Hartmanella sp.	0	0	0	*
Naegleria sp.	0	0	0	*
Arcella vulgaris	*	*	*	—
Centrophyxis aculeata	—	—	*	—
Euglypha sp.	0	0	—	—
Ciliates				
Trachelophyllum pusillum	0	*	*	0
Hemiophrys fusidens	0	*	*	*
Amphileptus calparedei	0	*	*	0
Litonotus fasciola	*	*	*	*
Trochilia minuta	*	0	*	—
Chilodonella uncinata	*	0	*	—
Cyclidium spp.	*	0	0	0
Glaucoma scintillans	*	0	0	0
Cinetochilum margaritaceum	*	0	*	—
Carchesium polypinum	*	*	*	*
Zoothamnium sp.	0	0	*	—
Vorticella campanula	0	0	0	—
Vorticella convallaria	*	*	*	*
Vorticella microstoma	*	*	*	*
Vorticella striata	*	*	0	—
Epistylis plicatilis	0	0	*	—
Opercularia spp.	*	*	*	—
Aspidisca cicada	0	*	*	0
Euplotes spp.	0	*	0	—
Stylonychia spp.	0	0	0	*
Tachysoma pellionella	0	0	0	—

[a] * = commonly found, 0 = recorded but not common, — = not recorded.

The same phenomena are found in RBCs where the discs near the inlet carry more flagellates and amoebae, followed by increasing numbers of bactivorous and carnivorous ciliates on subsequent discs. One would expect to

Table 5.2   Twelve most important protozoa in biological filters and activated-sludge plants in the United Kingdom (Curds, 1992).[a]

Ciliate species	Filter	Activated sludge	Food	Habit
Opercularia microdiscum	1	0	Bactivorous	Sedentary
Carchesium polypinum	2	7	Bactivorous	Sedentary
Vorticella convallaria	3	2	Bactivorous	Sedentary
Chilodonella uncinata	4	0	Filamentous organisms	Crawling
Opercularia coarctata	5	5	Bactivorous	Sedentary
Opercularia phryganeae	6	0	Bactivorous	Sedentary
Vorticella striata	7	0	Bactivorous	Sedentary
Aspidisca cicada	8	1	Bactivorous	Crawling
Vorticella microstoma	0	3	Bactivorous	Sedentary
Trachelophyllum pusillum	0	4	Carnivorous	Freeswimming
Vorticella alba	0	6	Bactivorous	Sedentary
Euplotes moebiusi	0	8	Bactivorous	Crawling

[a] Values range from 1, most important, to 8, least important and 0 indicates no importance.

find comparable changes in the protozoan populations in oxidation lagoons and perhaps also in constructed wetlands, although they have not yet been reported in the literature.

In activated-sludge processes, the situation is different because the microbial film carrying the protozoa moves along with the liquid. Furthermore, the settled sludge flocs with their attached protozoa are recirculated and mixed with fresh wastewater as it enters the biological reactor. In completely mixed systems, there is no opportunity for spatial distributions to occur, whereas in plug-flow systems the protozoan community changes as the wastewater flows along the reactor channel.

TEMPORAL SUCCESSION. Although spatial zonation of protozoa does not typically occur in activated-sludge processes, there is a succession of protozoan types during the maturation of activated sludge. Figure 5.4 shows an example of a typical succession.

Colorless flagellates are the first dominant protozoa to appear, typically represented by members of the euglenid and kinetoplastid groups. As the numbers of flagellates decrease, they are slowly replaced by free-swimming ciliates. These reach a peak after approximately 20 days and are replaced by crawling hypotrichous ciliates such as *Aspidisca* and *Euplotes* (Figures 5.2d, 5.2e, and 5.2f). Finally, these are replaced by attached peritrichs such as *Vorticella, Epistylis,* and *Opercularia* (Figure 5.3).

Such temporal successions have also been reported in RBCs, where flagellates and small amoebae colonize the bacterial biofilm within 1 day of startup. Rapid colonization by crawling and stalked ciliates then occurs and finally, carnivorous ciliates and large amoebae appear.

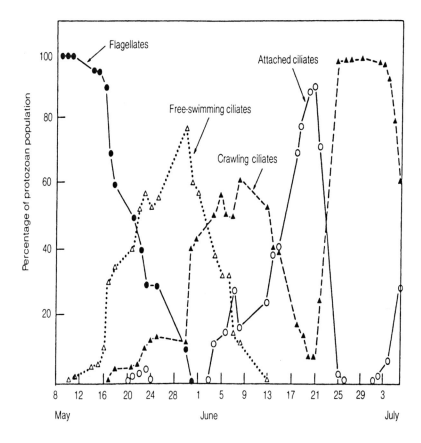

Figure 5.4   Successions of protozoa in an activated-sludge plant.

Several explanations for such successions have been suggested, more or less following those proposed for the zonal stratification in fixed-growth processes; that is, a combination of nutritional preferences and environmental conditions.

FACTORS AFFECTING PROTOZOAN POPULATIONS. In addition to the location within the unit process and the time (or duration) of treatment, there are other factors that affect the protozoan populations in WWTPs. These include various physical, biological, and chemical factors.

Dissolved Oxygen. Dissolved oxygen is one of the most important factors determining the presence or absence of a particular protozoan species in a wastewater treatment process. Protozoan species may be divided into those that must have free oxygen, those that will use it when it is present, and those that do not use it and are often unable to grow in its presence.

Trickling filters, activated-sludge units, and RBCs are typically operated in such a way as to keep the concentration of the mixed liquor at approximately

---

1 to 2 mg/L. However, even in aerobic processes, anaerobic conditions may prevail at certain stages. The bottom of an activated-sludge sedimentation tank, for example, is often anaerobic.

In constructed wetlands, the DO concentration varies according to the proximity to the plant roots; the interstitial fluid is well aerated close to the roots and becomes increasingly anaerobic further away from the roots. Consequently, a variety of protozoan types may be found with aerobic forms more abundant in the root zone and anaerobic forms, such as the ciliates *Metopus* and *Caenomorpha*, occurring elsewhere in the system.

pH and Carbon Dioxide Concentration. In nature, pH is intimately linked with both the carbon dioxide concentration and the local geochemistry (which influences the alkalinity of the water), and for this reason can be modified by the presence of living organisms. Chemical gradients, therefore, may exist around the sludge or biofilm. These gradients may play a part in holding a community of free-living protozoa together when associated with filamentous microorganisms. The greater pHs that are often present in oxidation lagoons may at least in part account for the lower abundances of protozoa compared to algae, yeasts, and fungi.

Carbon dioxide, the primary carbon source for the photosynthetic flagellates found at the top of a filter or oxidation lagoon, is toxic in high concentrations to many protozoa. It has been regarded as the primary adverse factor for limiting the distribution of ciliates. The ability to resist it is the distinctive feature of species associated with organically polluted environments.

Light. Light is the energy source for photosynthetic organisms and is quite limited in some wastewater treatment processes. Light does not penetrate more than a few millimeters in a trickling filter and the suspended solids in the mixed liquor of an activated-sludge unit prevent its penetration. The lack of light is a primary reason why so few photosynthetic protozoa are found in activated-sludge units and why they are confined to the upper layers in a biological filter.

Flow Rate. The wastewater flow rate affects not only the dilution rate (the reciprocal of the hydraulic residence time) of the plant, but also the velocity of the liquid passing over the surfaces of the treatment units. In the first case, greater flow rates will result in greater washout of free-swimming protozoa. Mean cell residence time (MCRT, reciprocal of sludge wastage rate) is similarly important in affecting those protozoa attached to the sludge flocs.

In the United Kingdom, the control of filter films is typically attributed to the grazing of macroinvertebrates, whereas in the United States, the scouring action of the liquid and the microbiological action are considered to be of greater importance. These differing views may result from the greater use (in the United States) of high-rate filters, with their attendant greater scouring action by the liquid. Therefore, flow rate may possibly affect protozoan populations by indirectly sloughing off the microbial film from the filters.

Toxic Waste. Toxic industrial wastewaters are often combined with domestic wastewater for treatment. Although there is considerable literature concerning the inhibition of various treatment processes by industrial wastes, comparatively little is known about their effects on protozoa. Heavy metals are probably the best studied toxics in this respect and several researchers have shown that heavy metals can affect the protozoan community structure within WWTPs (Gracia et al., 1994, and Madoni et al., 1996).

The concentrations at which various substances seriously inhibit the overall effectiveness of a biological treatment plant cannot be stated accurately, however, because toxicity is typically influenced by other environmental factors such as pH, DO, MCRT, other toxic substances, and temperature.

Predation. Wastewater treatment processes often contain predatory ciliated protozoa that typically prey on sessile peritrichs (i.e., peritrichs attached to a substratum by means of a stalk). Suctoria are also found and these feed on a wide range of protozoa. Predation is not limited to ciliate-eating ciliates; many of the hypotrichs ingest flagellates, while amoebae feed on both ciliates and flagellates. Little work has been carried out on the quantitative and kinetic aspects of protozoan predation.

Source of Organisms. If a habitat is continually inoculated with a supply of organisms from outside (i.e., from the wastewater or the atmosphere), provided the habitat is not immediately toxic, then these organisms will have an advantage over others not supplied from the outside. Land drainage and runoff provide a rich source of soil protozoa, particularly after a heavy rain. However, there is no quantitative information available on the number and species of protozoa carried in wastewater. Many protozoa are able to encyst and remain potentially viable for many years. Cysts may be transmitted in water, in the atmosphere, or even on the appendages of birds and flying insects.

Unit Process Operating Conditions. *TRICKLING FILTERS.* Trickling filters are designed to provide as great an internal surface area as possible. Protozoa that are capable of attaching directly to these surfaces have a clear advantage over those that cannot attach themselves because the latter will be subject to washout in the effluent before becoming established. Thus, a trickling filter will select for sedentary organisms. Because all of the protozoa that attach themselves to surfaces have a free-swimming stage in their life cycle, they too are subject to washout and greater wastewater flows will tend to increase the chance of reducing the protozoan populations and subsequent effluent quality.

ACTIVATED SLUDGE. Flow rate and the rate at which sludge is wasted can have a profound effect on the protozoan populations in the activated-sludge process. Any organism that settles and is returned back to the biological reactor after sedimentation has a much greater chance of survival than an organism that always remains in suspension because it is immediately subject to washout in the effluent. The only way a settling organism can be lost from

---

the system is when excess sludge is removed at intervals. Thus, the activated-sludge process selects for organisms that settle and peritrichous protozoa, such as *Vorticella, Epistylis,* and *Opercularia*, that attach themselves to the sludge floc are typically the dominant forms.

The retention time of the sedimentation tank is also of importance for the survival of protozoan populations. Microbial solids in the sedimentation tank may reduce the DO concentration. Low DO concentrations stimulate the formation of telotrochs, the free-swimming stage of attached ciliates, which then actively swim upwards towards greater DO concentrations and are thus washed out from the system. After some days, the ciliate population will change from one dominated by *Vorticella, Carchesium,* and *Epistylis* to one in which only *Opercularia* survives.

ROTATING BIOLOGICAL CONTACTORS. A correlation between organic loading and overall abundance of protozoa in RBC units with high or moderately high organic loadings has not been found. The number of protozoa and the number of peritrichs increases or remains the same throughout the units. In RBCs with low organic loading, total protozoan numbers and peritrichs typically decrease along the length of the units. The RBCs with high organic loading have few free-swimming ciliates.

PERFORMANCE MONITORING AND PROCESS CONTROL

It is evident that spatial and temporal distributions of protozoa and effluent quality are all interrelated. This has led to the proposal that protozoa can be used to indicate plant performance and effluent quality. In some cases effluent quality may be related to the presence or absence of certain protozoan species. This relationship occurs because while there is a fairly wide range of loading over which an individual protozoan species may live, there is a narrower range of loadings over which it may be present in large numbers. The dominant species present, therefore, can provide an immediate microbiological indication of the loading condition of the plant and of the quality of effluent expected.

It has been suggested that in wastewater treatment processes, the community structure of the protozoan population as a whole is a more reliable indicator of plant performance than numbers of individuals species or types. Specifically, in the activated-sludge process the numbers of organisms present has been shown to be irrelevant, to some extent, whereas the habit of the organisms present is of theoretical importance. There are also certain practical advantages to using community structure rather than indicator species for monitoring plant performance. If indicator species are used, it is essential that they be accurately identified, whereas in the case of community structures, identification of species is not always necessary. For example, it has been shown that in the activated-sludge process, grouping the protozoa by their trophic or behavioral characteristics makes observation easier for the nonspecialist without losing too much information. Similarly, it has been shown that certain ciliates that are either difficult to differentiate taxonomically or are known to have similar ecological roles may be grouped together for monitoring purposes. This has led to systems such as the Sludge Biotic Index (Madoni, 1994) for monitoring plant performance.

The processes best suited for protozoan-based monitoring systems are those that are easy to sample such as activated sludge, RBCs, and oxidation lagoons. Most of the work in this area has been carried out on the first two processes. So far, little attention has been paid to oxidation lagoons. By contrast, trickling filters and constructed wetlands (of the subsurface flow variety) are difficult to sample because of their construction. Taking a sample from either process typically involves digging it up, which is detrimental to its operation. Thus, there is limited opportunity for using protozoa for the routine monitoring of such systems.

ACTIVATED-SLUDGE UNITS. As the succession of various microbial populations in an activated-sludge unit develops, the quality of the effluent also improves. It is this link between effluent quality and the species of protozoa present that has led to the theory that protozoa are indicators of effluent quality. Curds (1992) cites a large-scale survey of treatment plants in the United Kingdom that indicated that there was a correlation between the protozoan species structure of the activated sludge and the quality of the effluent. The effluent from 56 plants was divided into four categories according to their 5-day biochemical oxygen demand (BOD_5) concentrations: very high quality (0 to 10 mg/L), high quality (11 to 20 mg/L), inferior quality (21 to 30 mg/L), and low quality (>30 mg/L). The frequencies of each species occurring in plants delivering effluents within each of these categories were calculated on a percentage basis (Table 5.3). Results indicated that the protozoa found within the mixed liquor were in some way associated with the effluent. It is evident from the table that *Carchesium polypinum*, for example,

Table 5.3 Percentage frequency of occurrence of some protozoa in plants producing effluents within the four ranges of BOD_5.

BOD$_5$ range, mg/L	Frequency of occurrence (%) and points awarded (in parenthesis)			
	0–10	11–12	21–30	>30
Vorticella convallaria	63 (3)	73 (4)	37 (2)	22 (1)
Vorticella fromenteli	38 (5)	33 (4)	12 (1)	0 (0)
Carchesium polypinum	19 (3)	47 (5)	12 (2)	0 (0)
Aspidisca cicada	75 (3)	80 (3)	50 (2)	56 (2)
Euplotes patella	38 (4)	25 (3)	24 (3)	0 (0)
Flagellated protozoa	0 (0)	0 (0)	37 (4)	45 (6)

was found principally in plants that produced good quality effluent, whereas flagellated protozoa were restricted to plants producing inferior effluents. From data such as these, a mathematical method has been devised for assessing effluent quality from an examination of protozoan communities. When tested at different treatment plants, 83% of the assessments made from this model were correct (Curds, 1992).

ROTATING BIOLOGICAL CONTACTORS. Rotating biological contactors are especially suited for development of indicator schemes based on protozoa and metazoa. Unlike most activated-sludge processes, RBCs are operated in a plug-flow mode and, therefore, have specialized protozoan and metazoan populations in each compartment. Additionally, in RBCs, biofilm from each compartment can be easily sampled.

Several indicators have been proposed for RBCs. Correlations between loading rate, biomass characteristics, and the predominance of various protozoa and metazoa have been found. At proper loading rates of 15 g $BOD_5/m^2 \cdot d$ or less, the ciliates *Carchesium, Epistylis, Vorticella, Opercularia, Cinetochilum*, and *Aspidisca* are the dominant organisms.

Effluent quality, as measured by BOD_5 and ammonia-nitrogen, has likewise been related to protozoan and metazoan groups. It has been shown that data for as few as 12 species of protozoa and metazoa may allow operators to predict effluent BOD_5 concentrations over three ranges: less than 25 mg/L, 25 to 40 mg/L, and greater than 40 mg/L (Kinner et al., 1988). Indicator schemes for RBCs must consider wastewater temperature and influent loading rates because some species seem to be seasonal.

ROLE OF PROTOZOA IN WASTEWATER TREATMENT PROCESSES

Although there is a large amount of literature pertaining to the role and importance of protozoa in the activated-sludge process, little work has been reported on the role of these organisms in other aerobic treatment processes. There is, however, great similarity between the types of organisms found in activated-sludge units, trickling filters, RBCs, and oxidation lagoons, and it is expected that they play a similar part in all four processes. The protozoa of constructed wetlands have yet to be studied in detail.

There are several ways in which protozoa affect effluent quality, the most important being the removal of bacteria by predation and enhancement of floc formation. Additional roles that protozoa play in the activated-sludge process include growth stimulation of bacteria and healthy flocs and degradation of organic waste. It is also possible that protozoa, and in particular the peritrich ciliates such as *Vorticella* and *Epistylis*, enhance nitrification.

REMOVAL OF BACTERIA. Bacterial numbers in wastewater are reduced during aerobic treatment. The dominant types of protozoa in these processes feed primarily on bacteria. Therefore, it has often been suggested that the predatory activities of the protozoa are responsible for bacterial removal.

Figure 5.5 summarizes the results of an experiment in which pilot-scale, activated-sludge processes were operated in the presence and absence of ciliated protozoa. Under protozoa-free conditions, highly turbid effluents of inferior quality were produced; most of the turbidity was a result of large numbers of bacteria suspended in the effluent. In the presence of protozoa, however, the clarity of the effluent was greatly improved and this was associated with a significant decrease in the concentration of viable bacteria in the effluent.

Quantitative studies suggest that, if protozoa in activated-sludge units feed at rates similar to those of the ciliate *Tetrahymena pyriformis* in pure culture (that is, consuming up to 500 bacteria per hour), then the protozoan populations typically found in these units could, by predation alone, easily remove sufficient quantities of bacteria to account for all of the observed reductions.

Protozoa are known to feed on a wide range of bacteria, including fecal coliforms such as *Escherichia coli* and various pathogenic bacteria, including those that cause diseases such as diptheria, cholera, typhus, and streptococcal infections. In activated-sludge plants it has been reported that, in the absence of protozoa, approximately 50% of *E. coli* entering in the wastewater are removed by unidentified processes, including natural die-off. When ciliates are present, approximately 95% of the *E. coli* are removed.

FLOCCULATION. Protozoa in pure culture are able to flocculate suspended particulate matter and bacteria. Furthermore, some protozoa, such as the amoebae, may have the ability to ingest flocculated bacteria. However, it is unlikely that protozoan-induced flocculation, or reduction in sludge production, are of significance in the activated-sludge process.

GROWTH STIMULATION OF BACTERIA AND HEALTHY FLOC.
The continual cropping of bacteria by protozoa not only removes excess suspended bacteria from the wastewater, but also seems to stimulate the growth of healthy bacteria, producing a more active floc. The purification of wastewater, as measured by the rate of oxygen consumption accompanying its degradation, occurs more rapidly and proceeds further in the presence of ciliates and bacteria than when bacteria alone are present. It is thought that predation by protozoa indirectly increases bacterial activity by preventing bacteria from reaching self-limiting numbers. Bacteria are thus kept in a state of prolonged physiological youth and their rate of assimilation of organic materials is greatly increased. Effluent quality, as measured by almost all typically used parameters, is thus improved by the activities of protozoa in the treatment process (Figure 5.5).

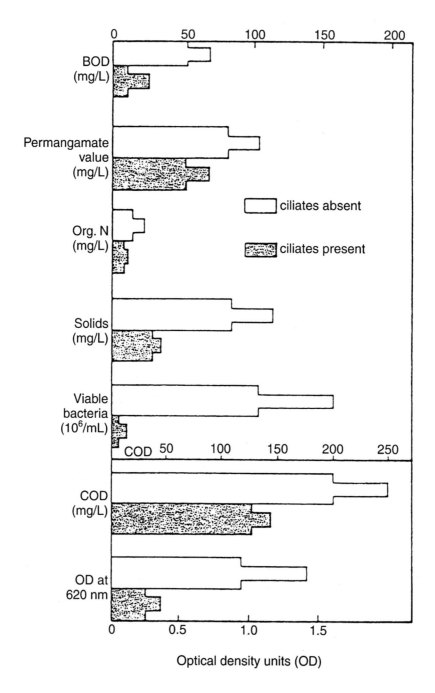

Figure 5.5 Bar chart of effluent quality issuing from laboratory-scale pilot plants operating in presence and absence of ciliated protozoa. The shoulders on the bars indicate ranges of variation between replicate plants (reproduced from Curds, 1992).

MICROSCOPIC EXAMINATION

Microscopic examination of the protozoa present may be used to provide an immediate microbiological indication of the loading condition of the plant and of the quality of effluent expected. Assessing protozoan populations in treatment plants, however, depends somewhat on sampling procedures. Because treatment plants are composed of various microhabitats that may support dissimilar aggregations of species, obtaining quantitatively comparable samples can be difficult. Two types of methods are typically used to sample for protozoa in wastewater treatment processes: direct sampling and the use of artificial solid support media.

DIRECT SAMPLING. Direct sampling involves extracting or examining naturally occurring material, such as biofilm layers or mixed liquor, in which the protozoa are present.

Trickling Filters. In the case of trickling filters, most protozoa are associated with biofilm layers attached to the filter bed media. The primary problem in taking samples is the limited access that operators have to other than the uppermost areas. In most cases, the only protozoa that can be readily examined are those associated with sloughed biofilm found in the effluent. Effluent samples should be taken as near to the base of the filter bed as possible. Protozoa in the effluent may be concentrated by passing the sample through a fine-mesh plankton net or by use of a centrifuge.

Rotating Biological Contactors. Protozoa in RBCs are also associated with biofilms. In this case there is no problem of access, and biofilm samples can be taken from any chosen area of a disc. The biofilm layers can then be examined qualitatively or quantitatively by direct microscope examination. In one method of quantitative analysis, the biofilm sample (36 cm^2) is placed in a graduated cylinder containing 5 mL of distilled water. The volume displaced by the sample is measured. The sample is homogenized by working it up and down in a series of Pasteur pipettes with progressively smaller tip diameters. The sample is then examined by one of two methods.

In the first method, a 20-μL aliquot of the sample is placed on a glass slide marked with 12 symmetrically placed dots. The sample is evenly distributed over an area of the slide that is covered by a cover slip. The fields directly below each dot on the slide are examined at 200X, and all protozoa present are identified and counted. The total number of species observed in the fields examined are multiplied by the dilution factor; this corrects for differences in thickness among biofilm samples.

A second method of examination may be used where the slides are scanned, making several horizontal passes per slide. This technique is preferred when motile species are present.

Some protozoa in the RBCs are also associated with the mixed liquor, and these may be analyzed quantitatively. Samples of the mixed liquor are

---

removed in 50-mL aliquots and centrifuged for 10 minutes at 2000 r/min. The residual pellet is placed in a graduated cylinder containing 5 mL of distilled water and the volume displaced is recorded. The sample is mixed using a Pasteur pipette technique and examined in the same way as described for biofilm.

Activated-Sludge Units. Most of the protozoa inhabiting activated-sludge units are associated with mixed liquor flocs. The protozoan populations may be analyzed qualitatively by extracting samples and examining them directly. Various methods for estimating the populations have been described. Use of counting chambers such as Sedgewick-Rafter slides and low-power stereo microscopes to count the numbers of protozoa in a small drop of known volume are common methods. Examining ten fields at 100X using a Z pattern is thought to give an optimum return. It has been shown that, using a Thomas counting chamber and a magnification of 100X, counting 5×10 µL of activated sludge is sufficient to recover 85% of ciliates and 78% of testate amoebae and to detect 94% of the ciliate species present (Augustin et al., 1989).

Another method that has been used involves homogenizing the sample by shaking it, and then mixing it with 2% xylocaine (0.5 mL) to yield a final volume of 10 mL. A drop of this mixture is transferred to a slide modified with 41 nylon threads arranged in parallel rows 1 mm apart. Organisms are counted, beginning at one end of the slide and continuing until the entire area in the 40 spaces created by the threads is scanned.

ARTIFICIAL SUPPORT MEDIA. Artificial support media are those that are introduced to a system and left for a period of time to allow colonization by protozoa. They are then withdrawn and the organisms collected are examined. A wide range of artificial solid support media have been used for collecting protozoa. Glass slides and polyurethane foam are typically used in WWTPs.

Glass slides provide an excellent support media for sessile protozoa and for those species closely associated with surfaces (the *Aufwuchs* community). In RBCs, glass slides may be attached to the discs and in activated-sludge units they are immersed in the biological reactors. The slides are left to allow colonization by protozoa. The optimal immersion time depends on physico-chemical and loading conditions and is typically between 3 and 9 days. The objective is to achieve maximum species diversity on the slide without allowing too much bacterial growth to accumulate and prevent observation. Following colonization, the slides can be examined directly under the micro-scope and for quantitative analysis, the number of individuals per unit area can be recorded.

For sampling free-swimming as well as *Aufwuchs* protozoa, polyurethane foam can be used. Polyurethane foam forms an open-cell structure that resembles an intricate latticework of pillars and interstitial spaces and permits rapid colonization by a wide variety of microorganisms. The primary disad-vantage of using polyurethane foam is that the samples cannot be examined

---

directly under the microscope. The protozoa must first be extracted by squeezing the foam and, because some protozoa are released more readily than others, this process could be selective for the group that easily releases.

Polyurethane foam has been used successfully for sampling protozoa in activated-sludge units. Small (1 cm^3) units of foam were left immersed in a biological reactor for 2, 3, and 10 days, and then squeezed to release the protozoa. Both types of artificial substrate (glass slides and polyurethane foam) yielded a wide range of protozoa (Sudzuki, 1981).

Examination of the isolated protozoa can be performed with a conventional microscope, although phase- and differential-interference contrast microscopy may permit easier identification. The rapid movement of protozoa may be retarded by the addition of a drop of 1% nickel sulphate solution. Likewise, lactophenol cotton blue also retards movement and reveals some structural details that are otherwise difficult to see.

*R*EFERENCES

Augustin, H.; Foissner, W.; and Bauer, R. (1989) Die Zahlung von Protozoan und Kleinen Metazoen inm Belebtschlamm. *Acta Hydrochim. Hydrobiol.* (Ger.), **17**, 375.

Corliss, J.O. (1994) An Interim Utilitarian ("User-Friendly") Hierarchical Classification and Characterization of the Protists. *Acta Protozool.* (Pol.), **33**, 1.

Curds, C.R. (1992) *Protozoa in the Water Industry.* Cambridge University Press, Cambridge, U.K.

Finlay, B.J.; Rogerson, A.; and Cowling, A. (1988) *A Beginners Guide to the Collection, Isolation, Cultivation and Identification of Freshwater Protozoa.* Culture Collection Algae Protozoa, Windermere, U.K.

Foissner, W.; and Berger, H. (1996) A User-Friendly Guide to the Ciliates (Protozoa, Ciliophora) Commonly Used by Hydrologists as Bioindicators in Rivers, Lakes, and Waste Waters, with Notes on Their Ecology. *Freshwater Biol.*, **35**, 375.

Gracia, M.-P.; Salvado, H.; Ruis, M.; and Amigo, J.-M. (1994) Effects of Copper on Ciliate Communities from Activated Sludge Plants. *Acta Protozool.* (Pol.), **33**, 219.

Hänel, K., (1979) Systematik und Okologie der Farblosen Flagellaten des Abwassers. *Arch. Protistenk.* (Ger.), **121**, 73.

Kinner, N.E.; Curds, C.R.; and Meeker, L.D. (1988) Protozoa and Metazoa as Indicators of Effluent Quality in Rotating Biological Contactors. *Water Sci. Technol.*, **20**, 199.

Lee, J.J.; Hutner, S.H.; and Bovee, E.C. (Eds.) (1985) *An Illustrated Guide to the Protozoa.* Soc. Protozool. Lawrence, Kans.

Madoni, P. (1994) A Sludge Biotic Index (SBI) for the Evaluation of the Biological Performance of Activated Sludge Plants Based on the Microfauna Analysis. *Water Res.* (G.B.), **28**, 67.

Madoni, P.; Davoli, D.; Gorbi, G.; and Vescovi, L. (1996) Toxic Effect of Heavy Metals on the Activated Sludge Protozoan Community. *Water Res.* (G.B.), **30**, 135.

Page, F.C. (1988) *A New Key to Freshwater and Soil Gymnamaoebae.* Culture Collection Algae Protozoa, Windermere, U.K.

Sudzuki, M. (1981) Faunistic and Ecological Studies of the Sewer Biota of Japan. *Verh. Int. Ver. Limnol.* (Ger.), **21**, 1094.

Chapter 6
Rotifers

INTRODUCTION

Rotifera is a group of aquatic organisms with approximately 2000 identified species. The name Rotifera comes from the Latin words *rota* meaning *wheel* and *ferre* meaning *bear*, hence wheel-bearing animal (rotifer). The wheels that rotifers have are actually disclike ciliated structures called corona and are the basic characteristic of the group. The corona is constantly moving, bringing food particles into the rotifer or in some cases acting as a means of locomotion.

Rotifers perform many beneficial roles in wastewater treatment. Their actions stimulate microfloral activity, enhancing oxygen penetration, and aid in the recycling of mineral nutrients, which helps stabilize organic waste. Unlike protozoa, they are rarely found in large numbers in wastewater treatment processes. Typically, there will only be one to four species of rotifer present in a well-operated wastewater treatment plant (WWTP).

DESCRIPTION

The phylum Rotifera is divided into two classes: Monogononta and Digononta (sometimes called Bdelloidea). The primary difference between the two classes is the number of gonads. Monogononta have one gonad each,

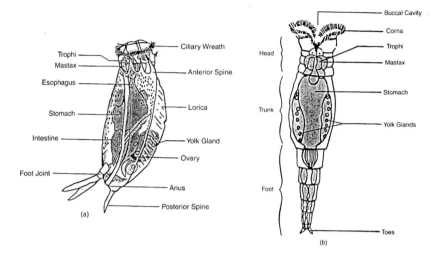

Figure 6.1   **(a) Typical Monogononta rotifer (modified from Stemberger, 1979) and (b) typical Digononta (bdelloid) rotifer (modified from Pennak, 1989).**

whereas Digononta have two (Figures 6.1a and 6.1b, respectively). Typical genera found in WWTPs are listed in Table 6.1.

Rotifers can be sac-shaped, spherical, or worm-shaped. They range in size from 40 to 500 μm and have an average life span of 15 to 45 days (Peljer et al., 1983). Except for the eyespot, most rotifers are colorless, although ingested food may sometimes give the organism the appearance of having

Table 6.1   List of typical rotifers found in wastewater.

Class Monogononta
  Order Ploima
    Family Brachionidae
      Subfamily Brachioninae
        Genus *Brachionus* spp.
        Genus *Euchlanis* spp.
        Genus *Kellicottia* spp.
        Genus *Keratella* spp.
        Genus *Monostyla* spp.
        Genus *Epiphanes* spp.
        Genus *Platylas* spp.
      Subfamily Colurin
        Genus *Lepadella* spp.
        Genus *Colurella* spp.
Class Digononta
  Order Bdelloida
    Family Philodinidae
      Genus *Philodina* spp.
      Genus *Rotaria* spp.

color. Reproduction in rotifers can either be sexual, asexual, or alternating. Members of the Class Digononta reproduce asexually; males are absent this class. Members of the Class Monogononta reproduce both sexually and asexually, although most of the time they reproduce asexually.

Rotifers have three distinct body regions; head, trunk, and foot (Figure 6.1). The head consists of the corona, buccal opening (mouth), and mastax. The mastax contains a set of jaws known as the trophus. The trophus is a hard body part found in the buccal cavity and may be one of several types. The function of the trophus is to filter out and obtain food particles. The type of trophus is an important feature for species identification (Pennak, 1989). The trunk houses the esophagus, stomach, intestinal track, gonads, and anus. The foot is a muscular structure that sometimes has toes and acts as a device to attach the rotifer to substrate. Some rotifers do not have either the foot or toes.

Some rotifers are covered with a cuticle layer called the lorica. This cuticle layer varies from having a few small plates to encapsulating the entire animal. In some species the layer is rigid and often an important characteristic for identification. The lorica may have spines at both the anterior and the posterior ends or have no spines at all (Figure 6.2).

Members of the class Digononta typically have well-developed corona, a cylindrical-shaped body that is capable of contracting and move with a wormlike creeping motion (Figure 6.3). Many members of this class can tolerate low oxygen environments and are often present in wastewater treatment systems.

The class Monogononta are saclike or spherical in shape and are characterized by the lorica covering the body, (Figure 6.1a). This class has males but they have a short life span and are rarely seen. When present their only function is to produce sperm and fertilize the females.

Because some rotifers feed on bacteria, detritus, and protozoa and others feed on phytoplankton or algae, changes in wastewater strength and composi-

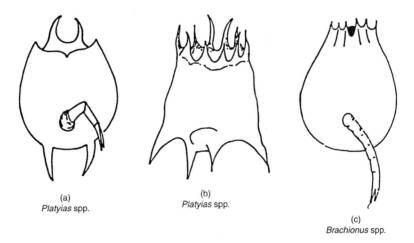

(a)
Platyias spp.

(b)
Platyias spp.

(c)
Brachionus spp.

Figure 6.2   Family Brachionidae: (a) *Platyias* spp., (b) *Platyias* spp., and (c) *Brachionus* spp. (modified from Pennak, 1989).

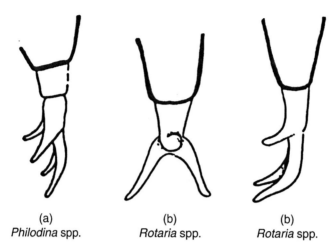

(a)	(b)	(b)
Philodina spp.	*Rotaria* spp.	*Rotaria* spp.

Figure 6.3 **Order Bdelloidea: (a)** *Philodina* **spp., (b)** *Rotaria* **spp., and (c)** *Rotaria* **spp. (modified from Pennak, 1989).**

tion may result in changes in the numbers and types of rotifers in the treatment process.

Rotifers typically display three types of feeding habits: vortex, where the corona creates water currents that bring food within the mouth; grasping, where the mastax captures the food; or trapping, where a wheel organ, the mouth, and pharynx combine to form a funnel to collect food.

Although rotifers typically occur in small numbers in wastewater treatment processes, they possess hearty appetites, which compensate for their numbers. For example, each *Brachionus calciflorus* in an oxidation lagoon is capable of consuming more than 12 000 phytoplankton cells per day (Klomowicz, 1968).

Rotifers move either by swimming or with a leechlike crawl. Free swimming is accomplished through the action of the ciliary currents produced by the head area or corona. The motion is forward and slow, with the direction of movement controlled by the foot, which acts like a rudder. Crawling is accomplished by a series of sequential movements that include:

(1) Extension of the body,
(2) Attachment of the anterior end of the rotifer to substratum by the adhesive glands of the retrocerebal sac,
(3) Contraction of the body to place the foot close to the head, and
(4) Attachment of the foot by the toes and adhesive glands to substratum.

These swimming and crawling motions permit the rotifers to increase in numbers in wastewater treatment process. Typically, Monogononta will predominate in lagoons, whereas Digononta will predominate in activated-sludge and fixed-growth processes.

In Monogononta, the degree of foot development varies. It may be reduced or absent altogether. These organisms are favored in lagoons where the ratio of free-water mass to substratum is high (when compared to activated-sludge and fixed-growth processes) and swimming is easily used.

Members of Digononta have a well-developed foot with up to four toes and adhesive glands or an attachment disk. Most of these organisms swim by the action of ciliary currents. The more active members, however, may move in a leechlike crawl. This crawling motion is frequently observed in activated-sludge and trickling filter systems where the ratio of free water to substratum is low (when compared to oxidation lagoons).

ROLE IN WASTEWATER TREATMENT

Rotifers perform many beneficial roles in stabilizing the organic wastes of lagoons and fixed-growth and activated-sludge processes. These include stimulating microfloral activity and decomposition, enhancing oxygen penetration, and recycling mineral nutrients.

In lagoons, rotifers feed predominantly on phytoplankton or algae. By continuously cropping the algae, the rotifers keep the algae population in a healthy state while recycling mineral nutrients. This action provides for a high level of wastewater treatment.

In fixed-growth processes, the primary role of the rotifers is to keep the biological slime layers in check by snipping off the top layer. Oxygen is able to penetrate to the lower layers and increased aeration and microbial stabilization of the organic wastes result. Keeping the slime layers porous also reduces ponding in trickling filters.

In the activated-sludge process, by consuming large quantities of bacteria, rotifers help keep the bacterial population healthy and in an active growing state. Additionally, the turbidity of the secondary effluent is reduced as the rotifers consume non-floc-forming bacteria.

Rotifers enhance floc formation by snipping off particles of activated-sludge floc to expose new absorptive surfaces for floc formation. This snipping action also permits oxygen to penetrate the floc. Mucous secreted at either the mouth opening or the foot also aids in floc formation and mucous-coated waste material excreted by the rotifers may act as a bonding site for bacteria.

In aerobic processes, rotifers' consumption of bacteria and solids contributes to biochemical oxygen demand (BOD) reduction. Rotifers living within the slime on the walls of biological reactors may also help to control slime growth and prevent anaerobic conditions within the slime layer (Calaway, 1968).

PROCESS CONTROL. The types and abundance of rotifers present in any wastewater treatment process are influenced by many factors, including dissolved oxygen, mean cell residence time, mixed liquor suspended solids, pH, temperature, and trophic level (BOD concentration). Because the rotifers present in the treatment process are in a state of ecological balance with their environment, any significant change in the environment may result in a change in the rotifer population.

Table 6.2   Simple illustrated key to some common rotifers found in wastewater treatment.

1. Single egg sack; hard shell (Figure 6.1a)—Class Monogonota

2. Paired egg sacks with soft body; body cylindrical and contractile; moves by creeping motion; Class Digononta, (Figure 6.1b)—Order Bdelloidea

3. Smooth cuticle; four toes: two on dorsal side, two on terminal end; short spurs (Figure 6.3a)—*Philodina* spp.
   Three toes: one dorsal, two terminal; two eye spots present (Figure 6.3b and c)—*Rotaria* spp.

4. Hard outer shell composed of two plates with spines; Order Plomia, Brachionidae, Brachioninae
   Hard outer shell present or absent and no spines

5. One very long posterior spine; four to six anterior spines (Figure 6.4a)—*Kellicottia* spp.
   Hard shell has polygonal pattern (Figure 6.4b)—*Keratella* spp.

6. Foot segmented (Figure 6.2a)—*Platyias* spp.
   Foot not segmented (Figure 6.2c)—*Brachionus* spp.
   With single long toe (Figure 6.5a)—Lecaninae, *Monostyla* spp.; (Figure 6.5b and c)—*Euchlanis* spp.; (Figure 6.5d)—*Epiphanes* spp.

7. Small rotifer; shell open at head. (Figure 6.6)—Brachionidae, Colurinae; last toe short (Figure 6.6a)—*Colurella* spp.; shell composed of dorsal and ventral plates (Figure 6.6b)—*Lepadella* spp.

The numbers, types, and activity (movement, feeding habits, etc.) of the rotifers present in the treatment process can be observed and related to process performance. Significant changes in rotifer populations may signal changes in wastewater strength and composition.

Typically, not only will there be one to four types of rotifers in any well-operated activated-sludge process, but each process will have its own characteristic group of rotifers. By determining the types and relative numbers of the organisms that are characteristic of the process when it is operating correctly, an operator will have a quick means of checking process conditions. For example, if a toxic substance enters an activated-sludge process, the rotifers would immediately show decreased activity or die. Typically, this condition may go undetected until there is total process failure. But if an operator notices a change in the rotifers present, it may be possible to limit the effect of the substance on the entire process.

Within a well-developed activated-sludge process, a succession of higher life forms occurs with increased stabilization of organic wastes (Gerardi, 1986). Rotifers are the next higher form of microlife beyond stalked ciliates. They are strict aerobes and are typically found in environments that contain at least 3 mg oxygen/L, but may also be found in wastewater with lower oxygen levels.

ROTIFER IDENTIFICATION

Daily, routine microscopic examination of the rotifers present in the waste-water treatment process can provide useful information. By correlating rotifer types, abundance, and activity to laboratory data, an operator can gain valuable insight to operation of the treatment plant. The organisms can be identified with an appropriate taxonomy key (Table 6.2).

To determine general shape and movement, rotifers must be viewed live. At low power (40X magnification), behavior patterns that are useful for identification are readily observed.

During observations, notes should be taken on the overall shape and size (length and width), number of egg sacks (single or paired), means of locomotion (attached or free swimming), and type of body (hard or soft). The rotifers present will be one of two types: illoricate (soft bodied) or loricate (hard bodied). The loricate rotifers are best identified when the soft part of the body has contracted leaving a clear view of the hard lorica. This helps when counting or observing the spines. A 10% solution of formalin is very effective for inducing contraction.

A different approach is required for soft-bodied rotifers. When under stress, these organisms contract, making detailed observation difficult. Because anatomical detail is important for identification, they should be viewed in as

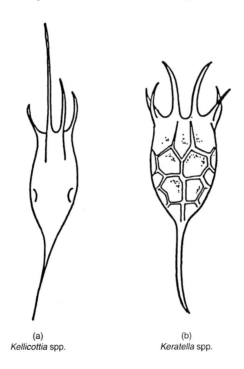

(a)
Kellicottia spp.

(b)
Keratella spp.

Figure 6.4   Family Brachionidae: (a) *Kellicottia* spp. and (b) *Keratella* spp. (modified from Pennak, 1989).

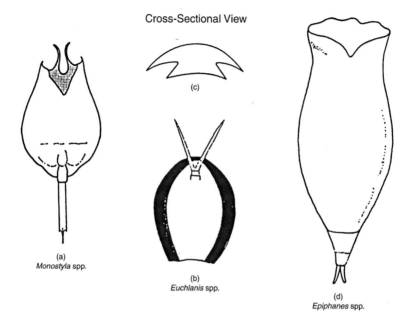

Cross-Sectional View

(c)

(a)
Monostyla spp.

(b)
Euchlanis spp.

(d)
Epiphanes spp.

Figure 6.5 **Family Brachionidae: (a)** *Monostyla* **spp., (b)** *Euchlanis* **spp.,
(c)** *Euchlanis* **spp. cross-sectional view, and (d)** *Epiphanes*
spp. (modified from Pennak, 1989).

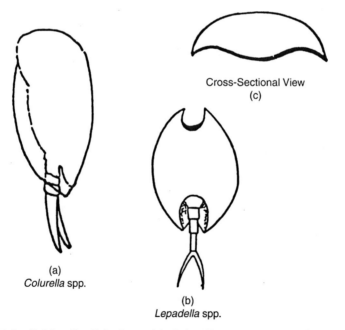

Cross-Sectional View
(c)

(a)
Colurella spp.

(b)
Lepadella spp.

Figure 6.6 **Subfamily Colurinae: (a)** *Colurella* **spp., (b)** *Lepadella* **spp.,
and (c)** *Lepadella* **spp. cross-sectional view (modified from
Pennak, 1989).**

natural a shape as possible. This is often difficult when the subject is moving around the microscope field of view. If the rotifers are moving too rapidly to view comfortably, a small drop of glycerine added to the slide will slow their movement.

SAMPLE COLLECTION AND PREPARATION

Once a sample is collected it should be taken to the laboratory immediately for examination. If the examination must be delayed, the sample should be refrigerated. There is no substitute for a fresh sample.

When preparing a sample for identification, permanent mounts can be made by placing a ridge of clear nail polish the size of a slip cover on a slide and letting it dry. This will form a partition and support the slip cover. A small drop of glycerine should be placed in the middle of the ridge formed by the nail polish, ensuring that air bubbles are not entrained. The rotifer is placed with a small drop of water onto the glycerine. A cover slip should be placed on top of the glycerin–water mixture and allowed to sit for 2 days or so until the water evaporates. Finally, a mounting material such as Canada balsam should be placed onto the specimen and sealed with nail polish. This slide can be stored and used for future identification.

Occasionally, the trophi must be viewed to identify genus and species. However, because of their location in the buccal cavity they cannot be viewed unless the soft body parts are dissolved.

The following method can be used for this process:

(1) Place the specimen on a clean slide and cover it with a slip.
(2) Place a drop of household bleach beside the cover slip.
(3) Using a small piece of tissue paper, touch the opposite side of the cover slip to draw the bleach under the cover slip and into the solution containing the rotifer.

After a few minutes the soft body parts will dissolve, leaving behind the trophi for examination.

DIFFERENTIAL COUNTS

When counting the number of rotifers in a sample, it is best to use a counting chamber such as a Sedgewick or Whitlock unit. These counting chambers are large slides with a partitioned-off area that holds a fixed volume of water. Because of the fixed volume, the count is always done on the same size sample volume.

The following steps should be followed to perform the count:

(1) Fill the counting chamber with sample and place a cover slip over the chamber.
(2) With a microscope on low power (40X), scan the chamber.
(3) Record the total number and types of rotifer present. For example, type A = 4, type B = 5, and so forth.
(4) Calculate a diversity index by dividing the number of each species by the total count.
(5) Compare laboratory results for a unit process with the diversity index and develop a range for expected process operation. It is helpful to plot the result on graph paper and monitor trends.

This technique determines the number of different rotifers present in the sample as opposed to the specific type of rotifer. This is a useful method when correlating treatment efficiency with rotifer types and diversity. If this is done over a long period of time and correlated with laboratory tests, it is possible to determine if the process is operating at an efficient level.

REFERENCES

Calaway, W.T. (1968) The Metazoa of Waste Treatment Processes—Rotifers. *J. Water Pollut. Control Fed.*, **40**, 412.

Gerardi, M. (1986) An Operator's Guide to Protozoa. *Pub. Works*, **7**, 44.

Klomowicz, H. (1968) Occurrence of Rotifers (Rotatoria) in Sewage Ponds. *Pol. Arch. Hydrobiol.*, No. 15.

Peljer, P.; Starkweather, R.; and Norgrady, T. (1983) *The Biology of Rotifers.* Dr. W. Junk Publishers.

Pennak, R.W. (1989) *Freshwater Invertebrates of the United States.* 3rd Ed., Wiley & Sons, New York.

Stemberger, R.S. (1979) *A Guide to Rotifers of the Laurentian Great Lakes.* EPA-600/4-79-021.

SUGGESTED READINGS

Arora, H.C. (1966) Rotifers as Indicators of Trophic Nature of Environments. *Hydrobiologia*, **27**, 146.

Gerardi, M.H. (1984) An Operator's Guide to Rotifers and Wastewater Treatment Processes. *Pub. Works*, **11**, 66.

Needham, J.G., and Needham, P.R. (1962) *A Guide to the Study of Freshwater Biology.* 5th Ed., Holden-Day, Inc., San Francisco, Calif.

Chapter 7
Nematodes and Other Metazoa

INTRODUCTION

The term metazoa refers to the members of the kingdom Animalia. Various representatives of this group can be found in wastewater treatment processes. The metazoa are multicellular organisms unlike the bacteria or protozoa, which are typically single cells. The cells in the metazoa seen in wastewater treatment processes are differentiated into tissues, which have various functions within the organism. These organisms are also heterotrophic, meaning they use organic material for food rather than manufacturing it through photosynthesis, as is the case with algae. The organic material they consume may consist of decaying organic matter, bacteria, protozoa, or other metazoans. The food is typically digested in an internal digestive tract, a cavity or tube where enzymes are secreted to break down ingested material. Metazoans also have unique embryonic stages and in many organisms a larval stage exists. Most grow slowly and, therefore, are found in older activated sludge or attached growth conditions. The term *very old sludge* may be a good one to describe their slow growth rate relative to the faster-growing bacteria and protozoa. Most seem to favor environments low in ammonia, so the systems containing metazoa are typically well nitrified.

The presence of multicellular organisms in wastewater treatment facilities is most likely the result of their presence in soil or freshwater environments and their entrance to the treatment process is probably caused by stormwater runoff infiltration in sewer lines or stormwater collected in combined sewer systems. Once within the treatment facility, conditions are often favorable for their proliferation and they have been found in significant numbers in various plants. Frequently they will proliferate for a time and then their numbers will decrease. It is often impossible to determine exactly where they came from or why they may dominate the plant population for a period of time, but undoubtedly they find the conditions uniquely favorable within the plant environment during those periods. Typically, they simply take their place as part of the natural food chain occurring among the organisms within the facility, feeding on other organisms in the population and reproducing.

The most often observed members of the kingdom Animalia in wastewater treatment facilities include nematodes (phylum Nematoda), water bears (phylum Tardigrada), bristle worms (phylum Annelida), gastrotrichs (phylum Gastrotricha), water fleas (phylum Cladocera), seed shrimp (phylum Ostracoda), and water mites (phylum Hydracarina). Others have also been observed, although less frequently, such as leeches (phylum Annelida). Each of these will be introduced in this chapter. Because nematodes are the most frequently observed of these, they will be dealt with in the most depth.

PHYLUM NEMATODA— NEMATODES (ROUNDWORMS)

CHARACTERISTICS. Free-living nematodes (Plates 1–3) are macroinvertebrates (animals lacking a backbone or spinal column) that are found virtually every place there is rotting organic matter, from the soil to lake bottoms. Their distribution is widespread and they have been found from the Arctic to the tropics in virtually every ecological niche. Wherever they are found their numbers can be enormous. Many have the ability to survive extremes of environmental conditions. Some can survive extremes of temperature and drying conditions by passing into a dormant (cryptobiotic) state until environmental conditions become more favorable.

Nematodes are important decomposers and are among the most numerous of all animals in terms of both number of species and individuals. Some nematodes are parasitic (living in the tissues of plants or animals) and consequently many of these have been studied in depth because of their immediate effect on the well-being of man. Those observed in wastewater treatment facilities are free living. Free-living nematodes consume bacteria, yeasts, fungi, and algae. They may be saprozoic (feeding on dissolved or decaying organic matter) or coprozoic (living in fecal material). Predatory species may consume rotifers, tardigrades, small annelids, or other nematodes. Although abundant, little is known about most free-living nematodes. They are numerous in soil, continuously entering wastewater treatment plants (WWTPs) and may be present in aerobic treatment processes in large, though variable numbers.

DESCRIPTION. Most species of nematode look much alike. The name nematode means *threadlike* in form. (The term *eelworm* is often used in the United Kingdom and it should be noted that this term may be encountered in literature on nematodes). They have a cylindrical shape and flexible cuticle, or outer layer. They are nonsegmented, but may appear to be segmented because of thickening of the cuticle. The cuticle is a tough, nonliving skin that is resistant to drying or crushing. Their bodies have no obvious external features and are cylindrical with tapering ends including a blunter anterior (head) end and a tapered tail. A mouth and lips are located at the anterior end and lead to the esophagus and digestive tract, which consists of a straight tube ending in the anus near the tip of the tail. Nematodes also have excretory, muscular, and reproductive systems. Most nematodes are less than 50 mm in length and many are microscopic in size. Although some parasitic nematodes are more than 1 m in length, free-living nematodes found in wastewater treatment facilities are microscopic in size, measuring 0.5 to 3.0 mm in length and 0.02 to 0.05 mm in width.

Movement is achieved through the action of longitudinal muscles against the internal pressure of the body fluids and the elastic properties of the cuticle. This form of locomotion allows the nematode to move through soil, mud, and tissues of plants and animals, and also permits the nematode to swim. Although a few swim by choice, free-living nematodes remain in contact with some substratum. Nematodes swim very inefficiently and do so in a whip-like or rapid thrashing motion, which is often observed in activated sludge.

REPRODUCTION AND GROWTH. The nematode sexes are typically separate, permitting sexual reproduction. During copulation, the spicules of the male extend into the vulva of the female. Fertilization occurs in the uterus and eggs are formed. The fertilized eggs, each containing a larva, are deposited in the soil or biomass of a WWTP. A fully developed juvenile, lacking only size and mature reproductive organs, hatches from each egg. Copulation, egg development, and larval hatching may take a few hours or a few weeks, depending on the nematode species and its environment. Reproduction in some nematodes is parthenogenetic (an embryo may develop from an unfertilized egg) and in others it is hermaphroditic (an individual functions as both the male and female producing both eggs and sperm). Those nematodes that are capable of hermaphroditic reproduction are often found in sand and gravel filter beds and belong in the genera *Mononchus* and *Ironus*.

The life cycle of free-living nematodes contains four larval stages (Figure 7.1). Each stage is followed by the molting of the cuticle, which permits the larva to grow in size while its sexual organs further develop. Maturity is achieved after the fourth and final molt. At maturity, the nematode is approximately 5 times its size at hatching.

CLASSIFICATION AND IDENTIFICATION. The most common groups of genera of free-living nematodes found in WWTPs belong to one or two superfamilies: Diplogasteroidea and Rhabditoidae. These two superfamilies can be identified and separated according to the structure of the esophagus as shown in Figure 7.2. Rhabditoidae has a median pharyngeal bulb without valves and a terminal bulb with valves. Diplogasteroidea is the opposite, possessing a median pharyngeal bulb with valves and a terminal bulb without valves.

Identification of nematodes is based largely on anatomical features, such as the shape of the mouth and esophagus, lip modifications, cuticle characteristics, and the shapes of the anterior and posterior ends. Identification should be made with an appropriate illustrated key because several different species may occur at the same time within a wastewater treatment system.

NEMATODE POPULATIONS WITHIN WASTEWATER TREATMENT PROCESSES. Nematodes thrive in aerobic habitats including wastewater treatment processes when dissolved oxygen (DO) concentrations are high and food is abundant. They may be found in large numbers in the effluent from WWTPs, particularly activated-sludge processes, because of their motility and resistance to chlorination. At present, no conclusive correlations have been found between the occurrence of the different genera of nematodes and specific aerobic treatment processes.

Aerobic treatment processes, particularly the activated-sludge and trickling filter processes, are continuously seeded with nematodes that are not removed

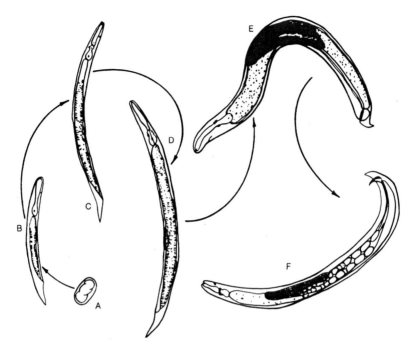

Figure 7.1 Life cycle of free-living nematodes: (a) egg, (b) first-stage juvenile, (c) second-stage juvenile following the first molt, (d) third-stage juvenile following the second molt, (e) fourth-stage juvenile following the third molt, and (f) adult following the fourth molt.

in the primary clarifier. Typically, the relative number of nematodes found in the activated-sludge process is significantly less than the number found in the trickling filter process. This is probably related to increased washout as a result of the suspended-growth nature of the activated-sludge process because they can more readily maintain contact with the zoogleal film that is present on the trickling filter media. Nematodes are present in relatively large and variable numbers in trickling filters and are often the most abundant macroinvertebrates. Because the biofilm of a trickling filter, unlike the floc of an activated-sludge unit, provides the nematodes with a stable substratum for attachment, rapid reproduction can occur. In addition, the residence time for the nematodes in the trickling filter is relatively large. These two factors often permit several complete generations of nematodes to grow within the same filter. The quiescent environment and relatively long residence time of the substratum or biomass of lagoon bottoms, rotating biological contactors, and sand filters are also favorable environments for nematode growth.

Factors Affecting Populations. In wastewater treatment processes, nematode populations are affected by several environmental factors, including parasitic fungi, DO concentrations, temperature and wastewater strength and composi-

---

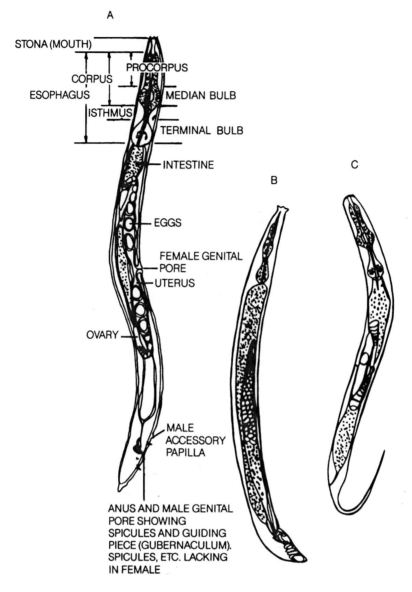

Figure 7.2 **Appearance and basic structures of common free-living nematodes found in aerobic wastewater treatment processes: (a) composite of various species, (b) rhabditid, and (c) diplogasterid.**

tion. Fungi, typical in wastewater, adversely affect the nematode population through parasitism.

Where sufficient moisture is available, some nematodes remain active at all temperatures between 0 and 47 °C (32 and 117 °F). Most nematodes also remain active at all pHs between 3.5 and 9. The range of temperature and pH tolerated by a particular species may, however, be quite narrow and therefore play a significant role in limiting its distribution and numbers.

Oxygen consumption in nematodes changes with growth and maturation. More oxygen is required by a nematode as it grows and matures; oxygen need is maximum for adult nematodes. Oxygen is supplied to the tissues in the nematode by diffusion. Therefore, in oxygen-poor environments, where sufficient oxygen cannot diffuse to the tissues in larger juveniles and adults, activity, growth, and reproduction are reduced. Under anaerobic conditions nematodes become quiescent. For this reason, they have been used as bioindicators of low DO or anaerobic environments. Warm wastewater temperatures cause decreased nematode numbers, whereas cold wastewater temperatures cause increased numbers. This is at least partially caused by the relationship between oxygen content of water and temperature; cold water is able to hold more DO than warm water.

Of the factors influencing nematode populations, changes in wastewater strength and composition have the greatest effect. Food seems to be the primary variable. Nematode numbers increase with greater wastewater strength and accompanying stabilization. Changes in wastewater strength that generate a significant change in the microflora or nematode diet may also generate a change in the types of nematodes present.

Role of Nematodes in Wastewater Treatment Operations. Nematodes perform several supportive roles in the overall biological stabilization of wastewater, sludges, and biosolids-amended soils. The most important are performed in the trickling filter process. Nematodes break loose portions of the biofilm coating the filter beds and, thus, prevent excessive growth of the biofilm and filter clogging. By burrowing or tunneling in the biofilm, nematodes keep the biofilm porous. Oxygen is able to penetrate the depths of the biofilm and increased aerobic microbial activity and degradation of organic wastes result. This burrowing action also recycles energy-rich compounds and nutrients, which further promote microbial activity. Bacterial growth is also stimulated by grazing nematodes that continuously crop the bacterial population in the biofilm.

Similarly, in sludges and biosolids-amended soils, the burrowing action permits greater diffusion of oxygen, increased microbial activity, recycling of nutrients, and cropping of the bacterial population.

MICROSCOPIC EXAMINATION. Nematodes can be viewed with a bright-field microscope at 100X magnification. This can be achieved with the 10X objective lens and the typical 10X ocular lens. Identification and classification can be confusing; however, this is not typically a problem because wastewater professionals are typically more interested in total numbers than in specific genera or species. Microscopic examination will reveal significant variations in shape and structure.

Before examination, the sample should be screened to remove particulate matter. Enumeration of the nematodes present can be made using a microscope counting chamber.

OTHER METAZOA

PHYLUM TARDIGRADA—WATER BEARS. Introduction. Most water bears (Figure 7.3 and Plate 4) are actually terrestrial organisms that live in the droplets and films of water that surround moist mosses and lichens. They are found in the capillary water surrounding sand grains near the water's edge on sandy beaches. Some species live in the bottom debris of freshwater water bodies and less than 10% are marine, living between sand grains in both deep and shallow seawater. There are between 300 and 400 known species of water bears worldwide.

Characteristics and Description. Tardigrades are mostly plant eaters, piercing algal filaments and moss cells with a stylet and sucking out the

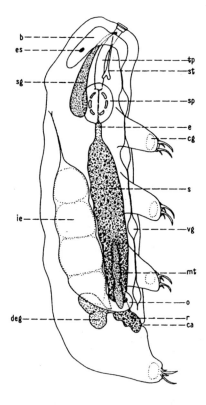

Figure 7.3   Semidiagrammatic lateral view of a typical tardigrade, 220X, with all musculature omitted. b: brain, ca: cloacal aperture, cg: claw gland, deg: dorsal excretory gland, e: esophagus, es: eyespot, ie: immature egg in saclike ovary, mt: Malpighian tubule, o: oviduct, r: rectum, s: stomach, sg: salivary gland, sp: sucking pharynx, st: stylet, tp: tubular pharynx, and vg: ventral ganglion (Pennak, 1989).

contents. Occasionally, however, the body fluids of small metazoans such as nematodes and rotifers are used as food and a small number are known to be carnivorous.

Mature individuals range from 0.05 to 1.2 mm in length, most being less than 0.5 mm long. They have a short, stout, cylindrical body and four pairs of stumpy, unjointed legs, which move in a pawing motion. Each leg is armed with four to eight claws used for clinging to the substrate. The head is simply the anterior portion of the trunk and contains a rather large brain. A pair of pigmented eyespots is present in most species. The esophagus empties into a midgut and fecal material is excreted through a cloacal opening or anus. The body is covered by a cuticle that is secreted and molted along with the claws four or more times in its lifetime.

Tardigrades have the ability to enter a state of suspended animation called cryptobiosis (formerly referred to as anabiosis) when environmental conditions become unfavorable, especially during periods of drying. During this period, metabolism slows considerably, the body becomes rounded and a considerable amount of water is lost. The time spent in this state before death occurs depends on the food reserves stored by the individual organism. The inactive state can persist for up to several years until environmental conditions become favorable again. While in a cryptobiotic state, tardigrades are highly resistant to harsh and abnormal environmental conditions. Revival is rapid if properly wetted.

Reproduction and Growth. Sexes are separate, but the vast majority of individuals are female. Depending on the species, the female lays between 2 and 30 eggs, which may be deposited singly or in groups. In some freshwater forms, the eggs are contained in the newly shed cuticle (Plate 5). These may sometimes be observed in the process of hatching within the shed cuticle. Free eggs are often sticky so they can attach to the substrate and tend to have a sculptured or textured surface, whereas eggs released into the old cuticle typically have a smooth surface. Eggs develop in 3 to 14 days, and the young emerge by rupturing the shell with their stylets. Newly hatched individuals are one-third to one-fifth the size of adults.

PHYLUM ANNELIDA—BRISTLE WORMS AND LEECHES.
Introduction. Annelids are segmented worms with a soft, muscular body wall covered by a thin cuticle. Lateral bristles (setae) increase traction with the substratum. Anterior eyespots are present in some genera.

The annelids are represented in the fresh water environment mostly by the Oligochaeta, aquatic earthworms, and Hirudinea, the leeches. The aquatic earthworms are most often seen in wastewater treatment facilities.

Bristle Worms. *CHARACTERISTICS AND DESCRIPTION.* The bristle worm (Figure 7.4) is the most frequently observed annelid in wastewater treatment facilities. It is an aquatic earthworm of the genera *Aelosoma* or *Chaena*. Bristle worms are typically less than 5 mm long. The segments are easily observed under a low power microscope, as are the bristles, or setae, associ-

---

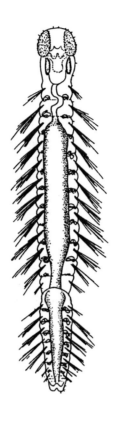

Figure 7.4 Dorsal view of *Aeolosoma*, a typical aquatic oligochaete showing separating bud, 50X (Pennak, 1989).

ated with each segment. Pigment globules give them a spotted appearance (Plate 6). The color of the pigment varies with the species, but is typically red or orange. In large concentrations they can cause *red sludge* or *pink foam* conditions.

Bristle worms feed on detritus and microorganisms by use of cilia on the ventral, or bottom, surface of the anterior end, or the prostomium. The prostomium is placed against the substratum and the central portion is elevated by muscle contraction, creating a partial vacuum, which dislodges food when swept into the mouth of the cilia. Movement is accomplished by undulation of the body and may be aided by beating of cilia on the anterior segment.

Some species of *Aeolosoma* are known to encyst when the water temperature is extremely cold. In this case a translucent, hardened coat of mucus is formed around the tightly coiled worm.

REPRODUCTION AND GROWTH. As with most aquatic species of annelids, bristle worms reproduce by budding—a new individual comes off the body of an older organism. The fission (budding) zone is located in one of the segments near the posterior end of the organism (Plate 7). The budding segment grows until four or more anterior segments of a new posterior worm and several new posterior segments of the anterior parent worm are formed. The two separate to become independent individuals. Under optimal growth conditions in activated sludge, *Aeolosoma* have been known to reproduce so rapidly that they turn settled sludge the color of the worm itself and cause excessively fast settling. They literally take over the system for unknown reasons. Systems west of Denver, Colorado, such as West Jefferson Sanitation District, continually fight *Aeolosoma* takeover. This has also been documented in Canada (WEF, 1995).

Leeches. *CHARACTERISTICS AND DESCRIPTION.* Approximately 44 freshwater species of leeches are known in the United States and they are common inhabitants of slow streams, ponds, marshes, and lakes, especially in northern regions. Although thought of as *blood-suckers*, only a few species will take blood from warm-blooded animals; most are predators and scavengers.

Mature leeches can range in length from less than 5 to 460 mm while swimming (the giant *Haemopis*). Leeches are flattened and segmented in form and the mouth is surrounded by an oral sucker. A caudal sucker is also present on the posterior end. They are highly muscular and the body shape can vary greatly. They can move along the substrate with a creeping or inchwormlike motion and many can swim rapidly using body undulations. Paired eyes are found at the anterior end.

Many leeches feed on dead animal matter or small invertebrates. It is for this reason that they can be occasionally observed in wastewater treatment facilities. They have been seen in amazing numbers in favorable environments high in organic pollutants. Oxygen exchange in freshwater leeches takes place through the skin. Little is known of their tolerance for low DO, but acidic environments are not favorable for them. Although rarely seen in wastewater treatment facilities, they have been observed in significant numbers in activated sludge.

REPRODUCTION AND GROWTH. Unlike other annelids, leeches do not reproduce asexually. All leeches are hermaphroditic. During copulation, the anterior end of one worm is directed toward the posterior end of the other. Eggs, contained with a nutrient rich cocoon, are laid from 2 days to many months following copulation. Most leeches have an annual cycle, breeding in the spring and maturing in the following year.

PHYLUM GASTROTRICHA—GASTROTRICHS. Introduction.
Gastrotrichs (Plate 8) are another type of microorganism typically found in freshwater environments on vegetation or in the debris of standing waters. Like water bears, they are typically found in water films of soil particles. Although not as abundant as rotifers, they are common inhabitants of ponds, streams, and lakes. They are found in puddles or marshes, especially in environments where there is detritus and decaying material. Consequently, the

DO content of these environments may be quite low, possibly less than 1.0 mg/L. Although their minimum oxygen requirement is not known, it is quite likely they may be able to withstand temporary anoxic conditions.

Characteristics and Description. Freshwater gastrotrichs (Figure 7.5) range in size from 0.07 to 0.6 mm in length, but most are between 0.1 and 0.3 mm long. They can range from short to long and wormlike, with the underside

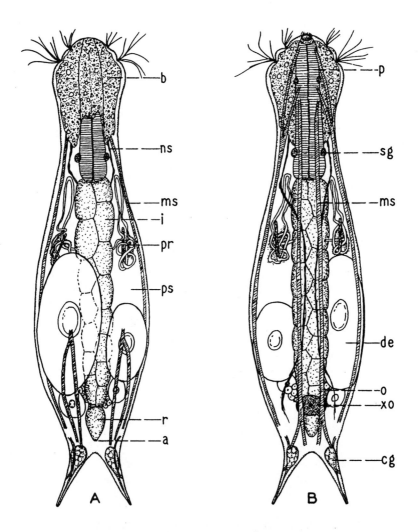

Figure 7.5 **Diagrammatic internal structure of a typical gastrotrich. A: dorsal, B: ventral, a: anus, b: brain, cg: cement gland, de: developing egg, i: intestine, ms: muscle strand, ns: longitudinal nerve strand, o: ovary, p: pharynx, pr: protonephridium, ps: pseudocoel, r: rectum, sg: salivary gland, and xo: "X organ" (Pennak, 1989).**

flattened. The head is typically separated from the body by a distinct neck area. Tufts of sensory cilia are typically seen on the head, sometimes between lobes, or slightly protruding areas. The mouth is in the front of or beneath the head and is often surrounded by short, delicate bristles. The digestive tract ends in an anus at the posterior end. Cilia under the head and trunk allow locomotion in a gliding manner along the substratum. Two toelike projections are typically present at the posterior end.

The body surface is covered by a cuticle, which in most species is patterned and textured with a variety of scales, and sometimes thicker plates. These may have small spines varying in length from quite short to rather long depending on the organism. The presence and size of the scales, plates and spines, and their characteristics, vary with the organism or location on the body. Most gastrotrichs temporarily leave the substrate and swim about, but their tendency is to remain near the substrate most of the time. The posterior end can be temporarily attached to a substrate by secretion of a cement-like substance. Muscle contraction allows for shortening, curving, and partially rolling up the body. Some species are able to spring or leap using long spines while flexing the body.

Food sources for the gastrotrichs vary. They can ingest bacteria, algae, small protozoa, and organic detritus. Food is brought into the mouth by currents created by beating cilia on the head.

Reproduction and Growth. Freshwater gastrotrichs are parthenogenic (embryos develop from an unfertilized egg), with only females known to exist. An individual typically produces one to five eggs during its lifetime. A mature egg is quite large and distends the body significantly before discharge. The egg is flexible as it is released and hardens with contact with the water. Sometimes the egg has spines or other protrusions. Eggs are typically attached to some small object on the substrate.

SUBPHYLUM CRUSTACEA, CLASS BRACHIOPODA, ORDER CLADOCERA—WATER FLEAS. Introduction. Cladocera (Plate 9) are common in nearly all freshwater habitats and are easily collected and observed because they are active swimmers. They include the genus *Daphnia* that is familiar to both amateur and professional biologists.

Characteristics and Description. Most members of the Cladocera (Figure 7.6) are between 0.2 and 3.0 mm long. In most species the central portion of the body is covered by a secreted shell or carapace, which is folded about the body and is open on the underside. The shell may appear oval, round, elongated, or angular. Various markings may be seen on the surface. In the groove along the under side are four to six pairs of trunk appendages, or small sets of legs. In many species, a spine is present on the posterior end.

A large, compound eye is obvious on the head. A beak (rostrum) projects on the front of the head. The head may appear helmet-like and the appearance of the head can vary seasonally. A set of very large antennae originates from the rear portion of the head. These antennae are powerful and used for

---

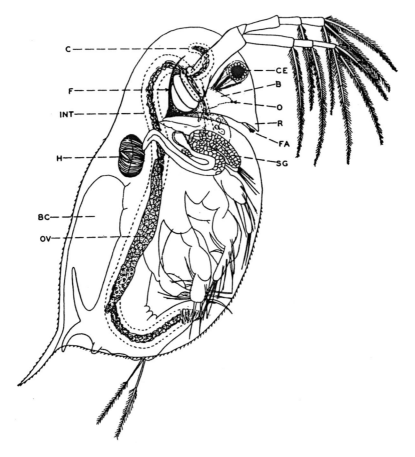

Figure 7.6 Anatomy of female *Daphnia pulex*, 70X, diagrammatic, muscles not shown. B: brain, BC: brood chamber, C: digestive caecum, CE: compound eye, F: fornix, FA: first antenna (antennule), H: heart, INT: intestine, O: ocellus, OV: ovary, R: rostrum or beak, and SG: shell gland (Pennak, 1989).

swimming, which occurs in a jerky, mostly vertical manner. The downstroke of the antennae move the animal upward and it slowly sinks using the antennae as a sort of parachute. Some species move in a smoother manner, beating the antennae or moving them in an oarlike manner. Some cladocerans can use the small trunk appendages for swimming and some are bottom dwellers, moving along the sediment.

Most of these organisms are filter feeders, using the trunk appendages to secure food and transfer it forward to the mouth. Their food consists of bacteria, algae, protozoa, and organic particles.

One curious trait causes problems in clarifiers. Some Cladocera rapidly move up and down through the water column. In activated-sludge secondary clarifiers this can cause significant mixing when a bloom occurs. This situation is often observed at Broomfield, Colorado, where Cladocera bloom approximately on the first of August each year. The bloom continues for a couple of weeks, and then disappears until the next year. During the bloom, effluent turbidity increases, apparently resulting from organism movement throughout the clarifier (Henderson, 1999).

Reproduction and Growth. During most of the year, in most habitats, reproduction is parthenogenic and female offspring are produced. The number of eggs produced at one time varies between 2 and 40 depending on the species and environmental conditions. Male specimens can be found in varying numbers in the spring, which leads to the sexual production of some eggs with females in the population capable of copulation. These are few in number compared to the eggs produced through parthenogenesis. There are four stages in the development of cladocerans: egg, juvenile, adolescent, and adult. Growth takes place through a series of molts.

SUBPHYLUM CRUSTACEA, CLASS OSTRACODA—SEED SHRIMP.

Introduction. Ostracods (Plates 10 and 11) are small crustaceans that are widely distributed in all types of freshwater and marine habitats. More than 2000 living species have been described. They are found in both standing and running waters. Ostracods are somewhat difficult to study because the body is enclosed in a bivalve shell resembling a clamshell (Figure 7.7). They resemble miniature mussels and have been referred to as *mussel shrimp* in European literature. This can lead to confusion with other organisms, so *seed shrimp* is the preferred term because without any magnification they resemble seeds.

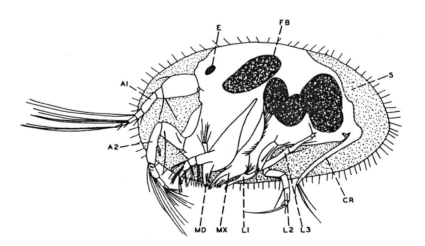

Figure 7.7 Female *Cypricercus reticulatus* (Zadd.), 80X, with left valve removed to show appendages, diagrammatic. A1: first antenna, A2: second antenna, CR: caudal ramus, E: eye, FB: food ball in digestive tract, L1: first leg, L2: second leg, L3: third leg, MD: mandible, MX: maxilla, and S: right valve of shell (Pennak, 1989).

Characteristics and Description. The freshwater species of ostracod found in the United States are typically less than 1 mm long and are seldom more than 3 mm in length. Colors range from white to brown, gray, red, brown, blackish, or even blotched. The bivalve shell is rounded or oval and contains calcium carbonate. A hinge line along the back is formed by a piece of cuticle. Muscles hold the valves together on the opening side. Inside the shell, the head region composes most of the body. Antennae protrude from the anterior end and are the primary appendages of locomotion. Trunk appendages, or legs, are small and may be used for swimming, feeding or clasping.

Most ostracods live near the bottom where they swim intermittently or creep or scurry over the mud and detritus. Their food consists mostly of bacteria, molds, algae, and decaying organic material. Some have been observed feeding on dead animals.

Reproduction and Growth. Reproduction is often parthenogenic, but may be sexual depending on the species. The spherical eggs are deposited singly or in clumps or rows on rocks, debris, or aquatic vegetation. Egg development is typically arrested during cold weather or unfavorable dry conditions. Eggs typically hatch in a few days to a few months and maturity is reached after a series of eight molts.

As with other metazoa, their growth within wastewater treatment systems is little understood. However, it is interesting that they characteristically grow in certain systems. For instance, the author has observed significant numbers of ostracods in extended aeration systems in Johnson County, Kansas.

CLASS ARCHNOIDEA, ORDER HYDRACARINA—WATER MITES.
Introduction. The water mites (Plate 12) are freshwater relatives of spiders, ticks, scorpions, and mites. They are found in almost all freshwater habitats, but they are more common in ponds and areas of lakes where aquatic vegetation is abundant. They are easily identified by their oval shape, bright colors, and clambering and swimming habits.

Characteristics and Description. Water mites (Figure 7.8) resemble spiders, but the cephalothorax (head region) and the abdomen are completely fused into a single unit. Most are soft-bodied, but some have a leathery or thickened cuticle forming a series of plates. Total length typically ranges between 0.4 and 3.0 mm.

As in other spiders, six pairs of appendages are always present, the last four pairs, legs, are the most obvious. Each leg is long and consists of six segments, which typically have some obvious spines or hairs, ending in two claws. The first two pairs of appendages are located on the head end by the mouth. Two pigmented eyes are located at the anterior end of the body. On the underside a genital area is located toward the posterior end.

Unlike other freshwater invertebrates, water mites have a striking coloration. Most are a shade of red or green, but many are blue, yellow, tan, or

---

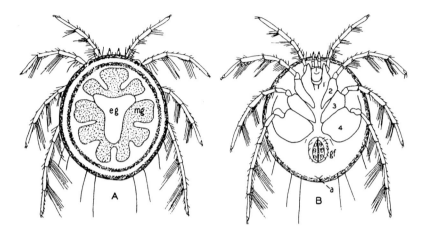

Figure 7.8   *Mideopsis orbicularis* (Müller) female, 40X. A: dorsal;
B: ventral; 1, 2, 3, and 4: first, second, third, and fourth
epimera, respectively; a: anus; eg: excretory gland;
gf: genital field; and mg: midgut (Pennak, 1989).

brown. The internal organs can give the appearance of markings. Some
species have more than one color displayed.

The Hydracarina swim with a rather uncoordinated flailing movement of
the legs. Some species creep along the bottom or among vegetation. They do
not stray far from the substrate. They tend to sink readily. The legs are
sometimes used for cleaning debris from themselves or one another.

Most water mites are carnivorous or parasitic, their food consisting of small
insects, worms, or host tissue. Some species feed on dead animals and some
have been observed to consume vegetation to an extent. Prey are held with the
legs and the bodily fluids are sucked out and the exoskeleton or indigestible
remnants discarded.

Reproduction and Growth. Reproduction in the Hydracarina is sexual. Eggs
are typically laid during the warm months. Eggs are typically red and are laid
in groups of 20 to 200 onto stones, vegetation, or debris. Eggs hatch in 1 to 6
weeks depending on the temperature and species. Larvae have mouths and
three pairs of legs. They are free swimming initially and then pass through a
parasitic period when they use aquatic insects as hosts. The larvae then
change into nymphs, which pass out of a pupae-like structure into the water.
The nymph is larger than the larvae and is carnivorous. The nymph stage
varies in length. Once again a pupaelike structure is formed and a mature
adult emerges.

Nematodes and Other Metazoa   87

METAZOA POPULATIONS WITHIN WASTEWATER TREATMENT PROCESSES

Most metazoans are found in close association with a substrate. They have traditionally been associated with activated sludge with longer mean cell residence times. There is some speculation that some actually dominate in the slime coatings found in association with the walls of the biological reactors and in other areas where the organisms would find suitable substrate. As these slime layers slough or are washed off, the organisms are washed into the sludge and can then be seen in the suspended growth environment of the mixed liquor.

All metazoans found in wastewater treatment facilities have found the environment to be suitable for their survival. They are believed to be washed into the system through infiltration and are able to survive and possibly reproduce in the facility. They simply take their place in the existing food chain and are observed by the operator.

REFERENCES

Henderson, N., Broomfield, Colorado (1999.) Personal Communication.
Water Environment Federation (1995) Earthworms in a Lively Plant. *Water Environ. Technol.*, **7**, 6, 14.

SUGGESTED READINGS

American Public Health Association; American Water Works Association; and Water Environment Federation (1998) *Standard Methods for the Examination of Water and Wastewater*. 20th Ed., Washington, D.C.

Calaway, W.T. (1963) Nematodes in Wastewater Treatment. *J. Water Pollut. Control Fed.*, **35**, 1006.

Chitwood, B.G., and Chitwood, Z.B. (1950) *An Introduction to Nematology, Section I: Anatomy*. Monumental Printing Co., Baltimore, Md., 213.

Gerardi, M.H. (1987) An Operator's Guide to Free-Living Nematodes in Wastewater Treatment. *Public Works*, **118**, 12, 47.

Goodey, T. (1963) *Soil and Freshwater Nematodes*. 2nd Ed., Wiley & Sons, New York, 544.

Nicholas, W.L. (1984) *The Biology of Free-Living Nematodes*. Clarendon Press, Oxford, U.K., 251.

Pennak, R.W. (1989) *Freshwater Invertebrates of the United States.* 3rd Ed., Wiley & Sons, New York.

Poinar, G.O., Jr. (1983) *The Natural History of Nematodes.* Prentice-Hall, Inc., Englewood Cliffs, N.J., 323.

Tarjan, A.C., et al. (1977) An Illustrated Key to Nematodes Found in Fresh Water. *J. Water Pollut. Control Fed.,* **49**, 2318.

Chapter 8
Filamentous Organisms

*I*NTRODUCTION

Filamentous organisms are typically observed in activated sludges. They are involved in a number of biochemical processes, including oxidation of carbonaceous compounds, sulfide, and iron. They also provide structural support for the integrity of activated-sludge flocs. Although their presence is desired, their excessive growth leads to operational problems known as filamentous bulking and foaming. Bulking is the term used to describe poorly settling and compacting activated sludge. Although activated-sludge bulking resulting from the presence of excessive amounts of filamentous organisms is most common, bulking in the absence or presence of low levels of filamentous organisms is occasionally observed and is referred to as *zoogleal bulking*, *viscous bulking,* or *slime bulking*.

*F*ILAMENTOUS ORGANISMS OBSERVED IN ACTIVATED SLUDGE

Studies have indicated that there are more than 30 different types of filamentous organisms present in activated sludge (Eikelboom, 1975; Eikelboom and van Buijsen, 1981; Jenkins et al., 1993; Scheff et al., 1984; Senghas and Lingens, 1985; Trick and Lingens, 1984; and Trick et al., 1984). However, not all of them are frequently observed. Some of them are reported only occasionally. The commonly observed filamentous organisms are shown in Table 8.1 (Blackbeard et al., 1986; Eikelboom, 1977; Richard et al., 1982; Strom and Jenkins, 1984; and Wagner, 1982). Some filamentous organisms are presented by their taxonomic names, for example, *Sphaerotilus natans*; others are designated by numbers such as type 1701 in the table. The reason for designation by a number is that their true taxonomic location cannot be established at the present time. As shown in Table 8.1, frequency of occurrence of filamentous organisms may vary with geographical location because of differences in wastewater characteristics, biological reactor design, plant operating conditions, and temperature. The most common filamentous organism observed in the United States has been *Nocardia* spp., followed by type 1701 and type 021N (Richard et al., 1982, and Strom and Jenkins, 1984). In the Netherlands, it has been *Microthrix parvicella,* type 021N, and *Haliscomenobacter hydrossis* (Eikelboom, 1977). Type 021N and type 0092 have been reported to be the most common filamentous organisms in Germany (Wagner, 1982) and South Africa (Blackbeard et al., 1986), respectively. In Australia, *M. parvicella* was the most common dominant filamentous organism, followed by type 0041/0675 (Seviour et al., 1994). *Microthrix parvicella* was also the most common in Italy, followed by *Nocardia* spp., type 0092, and type 0041 (Rossetti et al., 1994). Although more than 30 different types of filamentous organisms are observed in treatment plants, approximately 10 to 13 different types are responsible for most bulking and foaming episodes (Jenkins et al., 1993).

Table 8.1 Dominant filamentous organisms observed in bulking and foaming activated sludges.

Filamentous organism	Ranking in order of dominance			
	USA[a]	Netherlands[b,c]	Germany[d]	South Africa[e]
Nocardia spp.	1	—	5	7
Type 1701	2	5	8	8
Type 021N	3	2	1	—
Type 0041	4	6	3	6
Thiothrix spp.	5	19	—	10
Sphaerotilus natans	6	7	4	—
Microthrix parvicella	7	1	2	2
Type 0092	8	4	—	1
Haliscomenobacter hydrossis	9	3	6	—
Type 0675	10	—	—	4
Type 0803	11	9	10	—
Nostocoida limicola (types I, II, and III)	12	11	7	9
Type 1851	13	12	—	3
Type 0961	14	10	9	—
Type 0581	15	8	—	—
Beggiatoa spp.	16	18	—	—
Fungi	17	15	—	—
Type 0914	18	—	—	5

[a] Richard et al., 1982, and Strom and Jenkins, 1984: 525 samples from 270 treatment plants.
[b] Eikelboom, 1977: 1100 samples from 200 plants.
[c] *Nocardia* spp. not included in the survey. Only organisms causing bulking were covered by the survey.
[d] Wagner, 1982: 356 samples from 139 plants.
[e] Blackbeard et al., 1986.

EFFECT OF FILAMENTOUS ORGANISMS ON ACTIVATED-SLUDGE SETTLEABILITY

ACTIVATED-SLUDGE FLOC. To understand how filamentous organisms influence activated-sludge settleability, it is necessary to examine activated-sludge floc structure and its relation to filamentous organisms. An activated-sludge floc consists of bacteria, fungi, protozoa, some metazoa, and inorganic and organic particulates. Formation of activated-sludge flocs is accomplished by certain bacteria such as *Zoogloea, Pseudomonas, Achromobacter,*

Flavobacterium, Alcaligenes, Arthrobacter, and Citromonas (Jenkins et al., 1993). There are several hypotheses about how these organisms form flocs. However, the most accepted mechanism of floc formation involves the processes of microbial adhesion, biological flocculation, and aggregation. Floc-forming organisms produce substances called extracellular polymers that are various types of polysaccharide and glycoprotein fibers. These polymers surround the individual bacterial cell and enable the cell to stick to other cells or attach to other solid surfaces or aggregates, resulting in formation of flocs. In fact, extracellular polymers act similarly to synthetic polyelectrolytes and bring about floc formation by bridging and charge reduction. Other particles, inorganic or organic in nature, are also captured during the process of floc formation and may play an important role in the settleability of activated sludge. If the other particles are denser than the bacterial cells, the flocs settle faster.

Filamentous organisms are also incorporated to the flocs and play a significant role in providing structural support to activated-sludge flocs. Bacterial cells attach to filamentous organisms and form flocs around them. Therefore, filaments act as a backbone and make the floc stronger against breakup in the turbulent environment of a biological reactor. There might be three different types of flocs formed depending on the amount of filamentous organisms (Sezgin et al., 1978).

When filamentous organisms are mostly within flocs, strong flocs are produced. Such flocs settle fast, leaving a clear supernatant and resulting in a low sludge volume index (SVI) (Figure 8.1). If filamentous organisms are absent or very few, flocs produced may be weak and broken into smaller aggregates in the turbulent environment of the biological reactor. Such flocs are termed *pinpoint flocs*. In a settling test, larger flocs settle fast, resulting in a low SVI, and smaller, sheared particles settle slowly, resulting in a turbid supernatant. When excessive growth of filamentous organisms occurs, they extend from the floc surfaces and prevent formation of floc-to-floc dense aggregates. Instead, the aggregates formed are loose aggregates because of filament-to-filament or filament-to-floc bridging (Figure 8.2). Such aggregates settle slowly, resulting in a high SVI, but clear supernatant. Clear supernatant is caused by the filtering effect of filamentous organisms.

ACTIVATED-SLUDGE SETTLEABILITY. The settleability of activated sludge is evaluated by various parameters, including SVI. An SVI of 80 to 120 mL/g typically indicates a good-settling activated sludge and an SVI of greater than 150 mL/g may correlate to sludge bulking conditions. Because the SVI test is easily conducted, it is widely used by wastewater treatment plant (WWTP) personnel as an indication of settling characteristics of activated sludge. However, if test results are not carefully evaluated, they may lead to erroneous conclusions. The SVI test result is influenced by a number of factors, including suspended solids (SS) concentration, filamentous organism, and extracellular polymer levels. Although a high SVI caused by high levels of filamentous organisms or extracellular polymer levels indicates sludge bulking conditions, a high SVI because of high SS concentration may not indicate

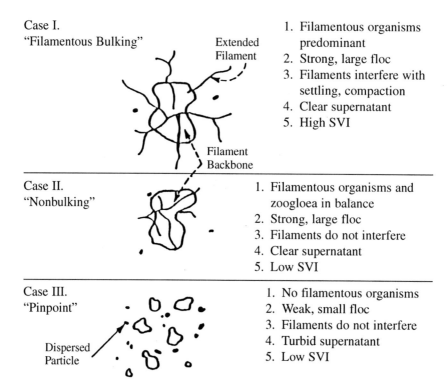

Case I.
"Filamentous Bulking"

Extended
Filament

Filament
Backbone

1. Filamentous organisms
   predominant
2. Strong, large floc
3. Filaments interfere with
   settling, compaction
4. Clear supernatant
5. High SVI

Case II.
"Nonbulking"

1. Filamentous organisms and
   zoogloea in balance
2. Strong, large floc
3. Filaments do not interfere
4. Clear supernatant
5. Low SVI

Case III.
"Pinpoint"

Dispersed
Particle

1. No filamentous organisms
2. Weak, small floc
3. Filaments do not interfere
4. Turbid supernatant
5. Low SVI

**Figure 8.1 Effect of filamentous organisms on formation of different types
of flocs in the activated-sludge process (Sezgin et al., 1978).**

sludge bulking but may lead to an inherent error in the test. Similarly, a low
SVI at relatively high SS concentration may not be an indication for good-
settling activated sludge because of the potential for hindered settling.

For a given sample, as the concentration of SS increases, SVI initially
remains constant. Further increases in SS concentration result in an increase
followed by a decrease in SVI (Figure 8.3). The flat section (constant SVI) of
the SVI profile in Figure 8.3 is a result of the fact that solids have completed
their zone settling and are undergoing compression settling by the end of the
30-minute settling period. The increase in SVI after the flat section with
increasing total suspended solids (TSS) concentration is a result of the
interference caused by bridging of aggregates during zone settling, resulting
in higher settled sludge volume. In this case, the incremental increase in
settled sludge volume occurs faster than the incremental increase of SS
concentration. This superficial increase in SVI is inherent to small settling
columns such as the 1-L graduated cylinder with which the SVI test is
conducted. The subsequent decrease in SVI with increasing TSS concentration occurs when the incremental increase in settled sludge volume is zero
(i.e., no settling) or less than the incremental increase in SS concentration.

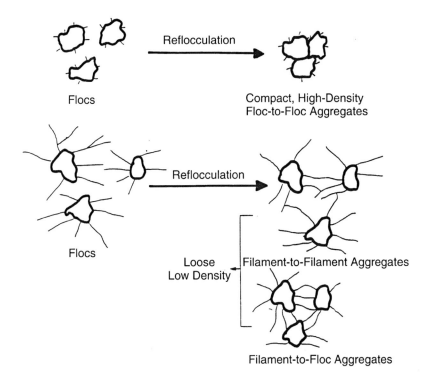

Figure 8.2 Effect of filamentous organisms on flocculation of activated-sludge flocs (Sezgin et al., 1978).

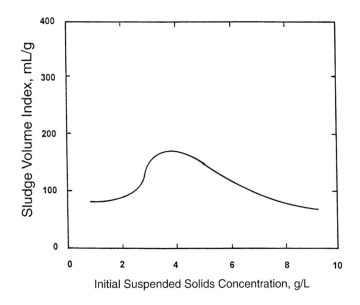

Figure 8.3 Variation of sludge volume index with initial suspended solids concentration.

To eliminate the effect of SS concentration on SVI, investigators introduced several variations of SVI such as diluted sludge volume index (DSVI), stirred specific volume index (SSVI), stirred specific volume index conducted at a TSS concentration of 3.5 g/L (SSVI$_{3.5}$), and SVI at constant TSS concentrations.

The DSVI was introduced by Stobbe in Germany in 1964 (Jenkins et al., 1993). The procedure is outlined in Table 8.2. The basis for this test is that the activated-sludge sample is diluted to the level that the resulting SS concentration has no influence on SVI.

The use of SSVI in place of SVI was proposed by White (1976). The procedure for conducting the SSVI test is similar to the SVI procedure outlined in *Standard Methods* (APHA et al., 1995). However, the test cylinder is different. White (1976) suggests that a cylinder with a diameter of 100 mm and a depth of 500 mm should be used for test. The volume of the cylinder is 4 L as opposed to 1 L as in the SVI test. White (1976) also recommends that the SSVI test be conducted at several TSS concentrations and the SSVI should be determined at a TSS concentration of 3.5 g/L by interpolating. The reason for reporting the SSVI at 3.5 g/L is partly because this concentration falls between the mixed liquor and return sludge TSS concentrations in most activated-sludge plants in the United Kingdom. Stirring speed used in the SSVI tests ranges between 1 and 4 r/min. The rationale for the use of a stirring device is that it aids in the reflocculation of activated sludge and eliminates the bridging among the flocs in small cylinders. However, slow stirring does not completely

Table 8.2 Diluted sludge volume index test (Jenkins et al., 1993).

Step	Action
1	Set up several 1-liter graduated cylinders
2	Prepare several 2-fold dilutions of the activated-sludge sample using either filtered or well clarified secondary effluent (i.e., no dilution, 1:1 dilution, 1:3 dilution).
3	Pour the activated sludges prepared in step 2 into graduated cylinders and mix the content of the cylinders to uniformly distribute the solids.
4	Let the activated sludges settle for 30 minutes and record the settled sludge volume.
5	Calculate the DSVI for the activated sludge resulting in a settled sludge volume which is less than, and closest to, 200 mL/L using the following formula:

$$DSVI \ (\text{mL/g}) \ = \ \frac{SV_{30} (\text{mL/L}) \times 2^n}{SS \ (\text{g/L})}$$

Where SV_{30} = 30-minute settled sludge volume, n = the number of 2-fold dilutions required to obtain settled sludge volume less than 200 mL/L, and SS = suspended solids concentration.

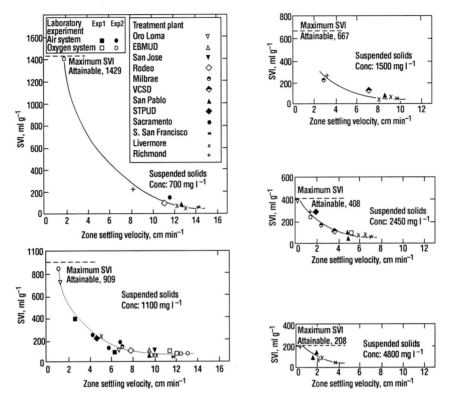

Figure 8.4   Correlation between sludge volume index and zone settling velocity (Sezgin, 1982).

eliminate the effect of SS on the test. Details of the stirring device are presented in *Standard Methods* (APHA et al., 1995).

The SVI test may also be conducted at a specified TSS concentration to evaluate the settleability of activated sludge. Sezgin (1982) reported on the correlation between SVI and zone settling velocity conducted at a SS concentration range of 0.7 to 4.8 g/L. The results are summarized in Figure 8.4.

Filamentous organisms have a significant effect on the settleability of activated sludge. Figure 8.5 shows the correlation between SVI and the total length of filamentous organisms extending from activated-sludge flocs grown in the laboratory on a domestic wastewater. The SVI is less than 100 mL/g when total extended filament length (TEFL) is less than 10^7 µm/mg TSS and increases sharply with increasing TEFL (Sezgin et al., 1978). Similar correlation has been reported between SVI tests conducted at a TSS concentration range of 0.7 to 4.8 g/L and TEFL for samples taken from 12 treatment plants (Figure 8.6) (Sezgin, 1982). The effect of filamentous organisms on activated-sludge settleability was also shown by other investigators. Lee et al. (1983) reported that DSVI and stirred SVI conducted at 2.5 g/L increased with increasing TEFL.

The effect of extracellular polymer concentration on activated-sludge settleability will be addressed under the section on viscous bulking.

Filamentous Organisms 99

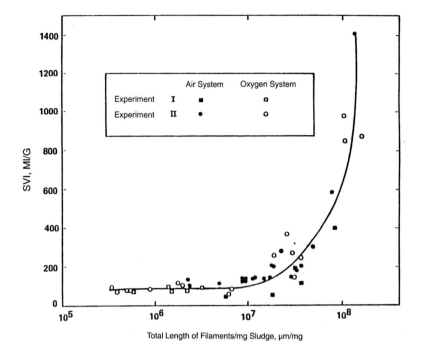

Figure 8.5   Relationship between sludge volume index and total length of filamentous organisms (Sezgin et al., 1978).

EFFECT OF FILAMENTOUS ORGANISMS ON SCUM FORMATION

Although most filamentous organisms cause bulking in activated-sludge processes, a few are known to cause scum on the surface of biological reactors and clarifiers (Sezgin and Karr, 1986). The filamentous organisms responsible for scum formation are Nocardioforms, *M. parvicella*, and type 1863 (Jenkins et al., 1993). Nocardioforms and *M. parvicella* have a tendency to float because of hydrophobic (no affinity for water) cell surfaces. When a sufficient number of *Nocardia* spp. grow within activated-sludge flocs, they also make flocs hydrophobic. The flocs then attach to air bubbles in the biological reactor and are carried to the liquid surface, forming a thick, brown scum layer. Scum on biological reactors and clarifiers causes severe operating problems in treatment plants and malicious odors in the vicinity (Jenkins et al., 1993, and Sezgin and Karr, 1986). Scum caused by *Nocardia* spp., *M. parvicella*, and type 1863 on secondary clarifiers may look similar in appearance to scum caused by denitrification. However, a microscopic examination

---

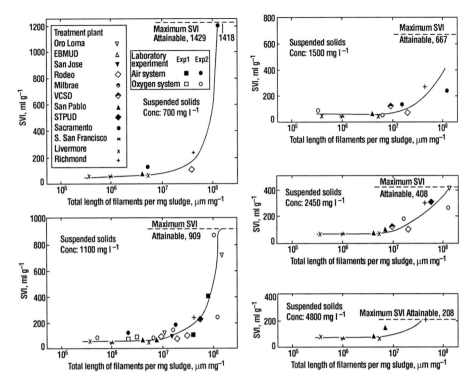

Figure 8.6 Relationship between sludge volume index and total length of filamentous organisms at different suspended solids concentrations (Sezgin, 1982).

will reveal the type of scum, whether it is *Nocardia*, *M. parvicella*, or type 1863, or caused by denitrification. Scum caused by denitrification looks similar in microbiological content to that of mixed liquor. On the other hand, scum caused by *Nocardia* spp., *M. parvicella*, and type 1863 contains excessive numbers of these organisms compared with the quantity of these organisms in the mixed liquor.

VISCOUS BULKING

In some cases, activated sludge exhibits poor settleability in the absence of filamentous organisms. Poor settleability may be caused by the production of excessive amounts of extracellular polymers, which are highly hydrophilic and retain large amounts of water. This retained water reduces the settleability of activated sludge. These types of settling problems are referred to as viscous bulking because of the slimy, jellylike appearance of the activated sludge (Hale and Garver, 1983). Floc-forming organisms produce extracellular polymers. In the past, species of the genus *Zoogloea* were presumed to be the only group of organisms capable of forming flocs and producing extracellular polymers. Therefore, viscous bulking is also referred as zoogleal bulking.

The presence of large amounts of extracellular polymers can be detected with either nigrosin or India ink reverse staining or the anthrone test for total sludge carbohydrates. The basis of the India ink test is that when a live sample of the activated sludge experiencing viscous bulking is stained, carbon particles of the ink cannot penetrate the thick slimy layer. Under a microscope, a black background with a clear layer of extracellular polymers surrounding the flocs is observed. The presence of large amounts of extracellular polymers can further be confirmed by the anthrone test. An activated sludge fed with domestic wastewater typically contains 14 to 18% carbohydrate expressed as glucose on a dry weight basis and an industrial activated sludge contains 20 to 22% (Jenkins et al., 1993). In severe cases, activated sludges may contain up to 70 to 80% carbohydrate. Settling problems caused by large amounts of extracellular polymer may start to occur at a carbohydrate concentration of 25 to 30% (Jenkins et al., 1983).

Viscous bulking may be caused by unbalanced growth of microorganisms. When an activated sludge is fed with a wastewater high in readily metabolizable, soluble organics and low in nutrients (nitrogen or phosphorus), organisms cannot produce nitrogen- and phosphorus-containing cell material. Instead, organisms produce excessive amounts of extracellular polymers from soluble organics (Jenkins et al., 1993). Overproduction of extracellular polymers has also been associated with high oxygen uptake rates mostly greater than 100 mg O_2/g VSS·h in plants treating domestic wastewater (Jenkins et al., 1993). In this case, unbalanced growth occurs because nutrients are not taken up as fast as the organic material is used. This type of problem has often been observed in domestic WWTPs using a high mixed liquor suspended solids (MLSS) concentration and in intensely aerated activated-sludge digestion processes (Jenkins et al., 1993).

Viscous bulking may lead to other operational problems. Extracellular polymers make activated sludge sticky and reduce its ability to flow. It has been reported that, under severe cases of viscous bulking, activated sludge would not flow into the return activated sludge (RAS) line (Jenkins et al., 1993, and Wanner, 1994). Viscous bulking can also cause significant solids dewatering problems.

MICROSCOPIC EXAMINATION

Information obtained from microscopic examination of the activated sludge is used to assess its condition, especially in relation to its separation in final clarifiers. Microscopic examination reveals information on floc morphology and size, types, and abundance of filamentous organisms, protozoa, and metazoa. An abundance of filamentous organisms may indicate poor settleability and compactability of activated sludge in final clarifiers or scum formation on the surface of biological reactors. Observation of small flocs may indicate turbid final effluent. Microscopic examination may also provide some information on influent wastewater characteristics (i.e., nutrient deficiency, high sulfide levels, presence of toxic substances, and types of waste), plant operational conditions (i.e., low dissolved oxygen [DO] concentrations, low food-to-microorganism ratio [F:M]), and influent wastewater feeding modes. It may further yield some information on biochemical activity of activated sludge (i.e., nutrient removal).

This section describes how to conduct a microscopic examination, which includes floc characterization, assessment of filament abundance, staining, and filamentous organism identification.

SAMPLE COLLECTION. Samples for microscopic examination of a mixed liquor should be taken below the surface so that foam or other floating material is not included. If foam is present on the surface of the mixed liquor, a sample of foam should also be taken to determine the presence and types of filamentous organisms. Foam samples should be taken from the surface of biological reactors, the surface of mixed liquor channels, and the surface of secondary clarifiers without the subsurface liquid. A sample size of 5 to 10 mL is sufficient for microscopic analysis. However, a larger sample can be collected to have a representative sample from the biological reactor.

In conventional activated-sludge processes, samples for mixed liquor should be collected at the location of good mixing either from the effluent end of biological reactors or mixed liquor channels before secondary clarifiers (Jenkins et al., 1993).

In activated-sludge processes using multiple reactors with different growth environments such as anaerobic–anoxic–aerobic reactors in biological phosphorus removal processes, it may be necessary to take samples from each reactor to observe certain floc characteristics, for example, presence of Neisser-positive staining cells (Jenkins et al., 1993). Similarly, in two-stage, activated-sludge processes in which the first stage is used for carbonaceous material removal and the second stage is for nitrification, it is necessary to take samples from each stage for microscopic examination (Jenkins et al., 1993). In some cases, a plant may use two parallel activated-sludge processes with little or no intermixing. In such cases, each process should be sampled separately for floc and filament characterization.

In some plants, certain filamentous organisms that may cause settling and compaction problems in activated-sludge processes may originate from other processes such as trickling filters or lagoons. The effluent from these processes should be sampled for filament characterization. Similarly, side streams to the activated-sludge process such as centrate from thickening centrifuges, supernatant from digesters, and centrate from dewatering processes should be sampled to determine the extent of filament contribution from these processes.

Frequency of microscopic examination depends on the circumstances at the plant. Daily microscopic examination is advised when the plant is experiencing bulking or bulking is anticipated, or RAS is being chlorinated or plant personnel are experimenting with different operational modes such as altering influent feeding points. Microscopic examinations for routine purposes may be conducted once per mean cell residence time (MCRT). More or less frequent observations may be established depending on frequency and severity of problems, seasonal variations in influent characteristics, and the budget for this activity.

SAMPLE TRANSPORT AND STORAGE. After collection, samples should be examined as soon as possible. If a delay in examination longer than several hours is inevitable, samples should be refrigerated at 4 °C. If samples are to be sent off site for analysis, ample air space should be left in the bottle to prevent the sample from becoming septic. For example, a minimum air space equal to sample volume should be maintained (Jenkins et al., 1993). Samples should neither be preserved with chemicals nor frozen.

Storage periods before microscopic analysis should not exceed 48 hours after sample collection and, if possible, samples should be examined within 24 hours. Because staining characteristics of samples may change with storage, it is best to prepare air-dried smears on microscopic slides at the time of sampling. A minimum of two slides should be prepared: one for Gram staining and the other for Neisser stain. The slides should be marked with sample location, date, and G for Gram staining and N for Neisser staining on the same side of the slide that the smear has been made.

EQUIPMENT AND SUPPLIES NEEDED FOR SAMPLE EXAMINA-TION. A phase-contrast microscope up to 1000X is required. The objectives with 10X and 100X are mostly used for examination; therefore, 20X and 40X objectives are useful, but not essential. Eye pieces with 10X should be used. One of the eye pieces should be equipped with an ocular scale to size flocs and filamentous organisms. The microscope should also have a built-in light source. The microscope should have a binocular head for convenient observation and a trinocular head if photography is desired. A 35-mm camera, a Polaroid instant film camera, or a digital camera can be used to take pictures for reference purposes. Daylight films can be used in photomicrography. However, the illumination needs to be adjusted with light-balancing filters; otherwise, the color will be affected. For more information on microscopy and photography, see Chapter 3, Chapter 12, and Appendix A.

Information on stains and staining procedures is presented in Figures 8.7 through 8.13. The most often used stains are Gram and Neisser stains. Other test methods such as sulfur oxidation and India ink tests to verify sulfide oxidation and detect presence of extracellular polymers, respectively, are provided. A staining procedure for intracellular storage products such as poly-β-hydroxybutyrate (PHB) and the crystal violet stain for sheath detection are provided. The anthrone test for quantifying the carbohydrate content of the activated sludge is also included.

SAMPLE EXAMINATION UNDER MICROSCOPE. The worksheets presented in Figures 8.14 and 8.15 are used as the basis for sample examination. On receipt of a sample, prepare thin smears of the sample on several microscopic slides. Place one drop of the well-mixed sample on a glass slide and spread evenly over approximately 50% of the slide surface. Label the

Reagents   Prepare every 3 to 6 months

Solution 1   Prepare the following separately, then mix **A** and **B**

A		**B**	
Crystal violet	2 g	Ammonium oxalate	0.8 g
Ethanol, 95%	20 mL	Distilled water	80 mL

Solution 2

Iodine	1 g
Potassium iodide	2 g
Distilled water	300 mL

Decolorizing Solution

Ethanol, 95%

Solution 3

Safranin O (2.5% w/v in 95% ethanol)	10 mL
Distilled water	100 mL

Procedure

1. Prepare thin smears of an activated-sludge sample on microscopic slides and thoroughly air dry. Do not heat dry.
2. Stain 1 minute with solution 1; rinse 1 second with water.
3. Stain 1 minute with solution 2; rinse well with water.
4. Hold slide at an angle and decolorize, with 95% ethanol added drop by drop to the smear for 25 seconds. Do not over decolorize. Blot dry.
5. Stain with solution 3 for 1 minute; rinse well with water and blot dry.
6. Examine under oil immersion at 1000X with direct illumination (not phase contrast): blue-violet is positive; red is negative.

Figure 8.7   Gram stain, modified Hucker method (Jenkins et al., 1993).

slide with sample location, date, and type of stain such as G for Gram, N for Neisser, and P for PHB. The slides can be stored indefinitely for staining later.

To prepare a wet mount for microscopic examination, follow the procedure given below:

(1) Place one drop (approximately 0.05 mL) of well-mixed activated sludge using a loop or a Pasteur pipette on a 25 mm × 75 mm microscope slide.

Reagents Prepare every 3 to 6 months

Solution 1 Prepare the following separately, then combine 2 parts by volume **A** with 1 part by volume **B**

A		B	
Methylene blue	0.1 g	Crystal violet	
Ethanol, 95%	5 mL	(10% w/v in 95% ethanol)	3.3 mL
Acetic acid, glacial	5 mL	Ethanol, 95%	6.7 mL
Distilled water	100 mL	Distilled water	100 mL

Solution 2

Bismark brown, $C_{18}N_{18}N_8$ (1% w/v aqueous)	33.3 mL
Distilled water	66.7 mL

Procedure

1. Prepare thin smears of an activated-sludge sample on microscope slides and thoroughly air dry. Do not heat dry.
2. Stain 30 seconds with solution 1; rinse 1 second with water.
3. Stain 1 minute with solution 2; rinse well with water; blot dry.
4. Examine under oil immersion at 1000X with direct illumination (not phase contrast): blue-violet is positive (either entire cell or intracellular granules); yellow-brown is negative.

Figure 8.8 Neisser stain (Eikelboom and Van Buijsen, 1981, and Jenkins et al., 1993).

(2) Place a 22-mm No.1 glass coverslip on the drop. If the coverslip floats on the sample, it indicates that too much sample was placed on the slide and the filaments cannot be examined at high magnifications properly. In this case, either a new slide should be prepared or the coverslip should be pressed down gently with a blunt object to expel the excess water. Then, the excess water can be removed from the sides of the slide using a tissue.

(3) Examine the wet mount at 100X under phase-contrast illumination.

General Characteristics of Sample. Summarize the observations in the worksheets provided in Figures 8.14 and 8.15. The summary should include the following characteristics:

- Size and morphology and color of flocs. Measure approximately 10 to 20 flocs in three arbitrary fields and record the number of flocs in the following categories:
  — Floc size categories based on maximum floc dimension: small, less than 150 μm; medium, 150 to 500 μm; and large, bigger than 500 μm.

Test A

Prepare fresh weekly

Solution

Sodium sulfide ($Na_2S \cdot 9H_2O$)   0.1 g
Distilled water   100 mL

Procedure

1. Mix one drop of an activated-sludge sample with one drop of sodium sulfide solution on a microscopic slide.
2. Let the mixture stand open to the air for 10 to 20 minutes.
3. Place a coverslip on the mixture and gently press; remove expelled solution with a tissue.
4. Examine under oil immersion at 1000X with phase contrast. The observation of highly refractive, yellow-colored intracellular granules (sulfur granules) indicates a positive test.

This test may yield variable results, then an alternative sulfur oxidation test (test B) may be run

Test B

Prepare fresh weekly

Solution

Sodium thiosulfate ($Na_2S_2O_3 \cdot 5H_2O$)   0.1 g
Distilled water   100 mL

Procedure

1. Let biomass in an activated-sluge sample settle and transfer 20 mL of supernatant to a 100-mL Erlenmeyer flask.
2. Mix the activated-sludge sample and add 1 to 2 mL of the sample to the flask.
3. Add 1 mL of sodium thiosulfate solution to the flask.
4. Place the flask on a shaker and shake it overnight at room temperature.
5. Examine under oil immersion at 1000X with phase contrast as in test A.

Figure 8.9   Sulfur oxidation test (Jenkins et al., 1993).

Reagent

India ink

Procedure

1. Mix one drop of India ink with one drop of an activated-sludge sample on a microscope slide. The sample volume may need to be adjusted depending on the ink used.
2. Place a coverslip on the mixture and examine at 100X using phase contrast.
3. In activated sludge containing normal levels of exocellular polymeric material, the carbon particles in the India ink solution penetrate the flocs almost completely, at most leaving a clear center.
4. In activated sludge with large amounts of exocellular polymeric material, large, clear areas containing a low density of cells are observed.

Figure 8.10   India ink reverse stain (Jenkins et al., 1993).

Solution 1

Sudan Black B (IV), 0.3% w/v in 60% ethanol

Solution 2

Safranin O, 0.5% w/v aqueous solution

Procedure

1. Prepare thin smears of an activated-sludge sample on microscope slides and thoroughly air dry.
2. Stain 10 minutes with solution 1; add more stain if the slide starts to dry out.
3. Rinse 1 second with water.
4. Stain 10 minutes with solution 2; rinse well with water; blot dry.
5. Observe under oil immersion at 1000X with direct illumination (not phase contrast. Poly-β-hydroxybutyrate granules will appear as blue-black granules within the cell, while the cytoplasm will be pink or clear.

Figure 8.11   Poly-β-hydroxybutyrate stain (Jenkins et al., 1993).

Reagent

Crystal violet, 0.1% w/v aqueous solution

Procedure

1. Mix one drop of an activated-sludge sample with one drop of crystal violet solution on a microscope slide.
2. Place a coverslip and observe at 1000X using direct illumination. Cells will stain deep violet; the sheaths will appear clear to pink in color.

Figure 8.12   Crystal violet sheath stain (Jenkins et al., 1993).

Reagents

Sulfuric acid   Mix 750 mL concentrated sulfuric acid (75% v/v reagent grade) with 250 mL distilled water (add the acid to the water). Cool the mixture to room temperature before using.

Anthrone reagent Add 200 mg anthrone (9 [10H]-anthracenone; $C_6H_4COC_6H_4CH_2$) to 5mL absolute ethanol and dilute to 100 mL with 75% sulfuric acid. Store at 5 °C and prepare monthly or sooner if a brown color is observed.

Glucose standard Dissolve 100 mg glucose in 100 mL distilled water and add 150 mg benzoic acid. Store at 5 °C. Dilute 1:10 ration with distilled water.

Sodium chloride   Add 8.5 g NaCl to 1000 mL distilled water (0.85% solution).

Equipment   Pyrex boiling tubes (thick-walled, 15 cm × 2.5 cm)
Boiling water bath
Ice waterbath
Spectrophotometer (625 nm) and cuvettes

Procedure

1. Centrifuge an activated-sludge sample with a known suspended solids concentration; discard supernatant and wash the solids with 0.85% NaCl solution; recentrifuge; resuspend to original volume with distilled water. Omit this step if washed and unwashed samples yield the same results.
2. Pipette samples ranging from 0.1 to 1.0 mL in volume into a series of boiling tubes; adjust all final volumes to 1.0 mL with distilled water.
3. Prepare three to four glucose standards in the range of 10 to 100 µg glucose/mL distilled water using the 100-µmg/mL glucose standard (for example, dilute 0.1 mL of 100-µg/mL glucose standard to 1.0 mL distilled water to make a 10-µm/mL glucose standard). Also, include 1.0 mL distilled water blank.
4. Place all the tubes in the ice waterbath.
5. Add 5.0 mL chilled anthrone reagent to each tube and mix the content of the tubes well and leave them in the ice waterbath.
6. Transfer all tubes to the boiling water bath and keep them in the boiling water for exactly 10 minutes.
7. Transfer the tubes back to the ice waterbath.
8. Measure the absorbance of all tubes at 625 nm and use a tube filled with distilled water as a blank.
9. Plot log percent transmittance versus glucose concentration using the transmittance of the tubes containing glucose in the range of 10 to 100 µg glucose. The plot should be a straight line. Determine the glucose content of samples from the plot using their transmittance values. Use only sample transmittance values that are between ≤90% transmittance and the percent transmittance produced by the highest glucose standard that lies on the straight line portion of the plot. The results are obtained as micrograms per milliliter glucose. Convert this concentration to milligrams total carbohydrate per gram dry weight activated sludge or as percent of dry weight as glucose.

Figure 8.13   Anthrone test for carbohydrate (Jenkins et al., 1993).

Sample Number_____   Sample Location _____

Sample Date ___/____/_____   Observation Date ___/___/_____

Filament Abundance:

☐   ☐   ☐   ☐   ☐   ☐   ☐

| 0 | 1 | 2 | 3 | 4 | 5 | 6 |
| None | Few | Some | Common | Very Common | Abundant | Excessive |

Filament Effect on Flock Structure: ☐   ☐   ☐

Little or None   Bridging   Open Flock Structure

Morphology of Floc: ☐   ☐   ☐   ☐   ☐   ☐

Firm   Weak   Round   Irregular   Compact   Diffuse

Floc Color: ☐   ☐   ☐

Light Brown   Golden Brown   Gray or Black

Flock Size (μm):
(% in range)

<150	150–500	>500

Features:

Free cells in suspension _____   Neisser-positive cell clumps _____

Zoogloes _____   India ink test _____

Spirochaetes _____   Chlorine damage to filaments _____

Inorganic/organic particles _____

Filamentous Microorganism Summary:

	Rank	Abundance		Rank	Abundance
Nocardia spp.			*M. parvicella*		
Type 1701			Type 0581		
S. natans			Type 0092		
Type 021N			Type 0803		
Thiothrix spp.			Type 1851		
Type 0041			Type 0961		
H. hydrossis			Type 0675		
N. limicola			Other		
Fungus			Other		

Remarks_____

_____

_____

Figure 8.14   **Worksheet for microscopic examination of flocs and filamentous organisms.**

Sample Number_____   Sample Location _____

Sample Date ___/___/_____   Observation Date ___/___/_____

Comments:

Observation of Protozoa and Metazoa:

Protozoa	Abundance	Metazoa	Abundance
Amoebae		Rotifers	
Flagellates		Nematodes	
Crawling ciliates		Annelids	
Free-swimming ciliates			
Stalked ciliates			

Wet Mount Observation; 1000X, Phase Contrast for Filament Characteristics
1000X, Bright Field for Stains

Filament No.	A	B	C	D	E
Branching					
Motility					
Filament Shape					
Filament Location					
Attached Growth					
Sheath					
Crosswalls					
Filament Diameter, µm					
Filament Length, µm					
Cell Shape					
Cell Size, µm					
Sulfur Deposits					
Other Granules					
Gram Stain					
Neisser Stain					
Abundance					
Rank					
Identification					

Figure 8.15   Worksheet for observation of protozoa and metazoa and identification of filamentous organisms.

- Percentages of flocs in each category: for example, if 20 flocs are sized in each of three fields with the following results: 20 small, 30 medium, and 10 large flocs, the percentages of flocs in each floc size category are 33% small, 50% medium, and 17% large.
    - Morphology: round and compact or irregular and diffuse.
    - Color: light brown may indicate low MCRT operation, and gray or black color may indicate septic influent or anaerobic conditions in the biological reactor. Golden brown floc color may be the result of good operating conditions, including proper DO and moderate MCRT.
- Presence of free cells in the sample. If the supernatant of the settled activated sludge is turbid, it is an indication that free cells are present in the sample. Dispersed growth (free cells) indicates several conditions: low MCRT (or high organic loading), low DO, low pH, presence of toxicants or surfactants, and mechanical shearing and growth of certain types of organisms (i.e., type 1863, *N. limicola*, *H. hydrossis*, or the Neisser-positive tetrads).
- Presence or absence of zoogloeal colonies. These organisms can be in amorphous or fingered forms and indicative of low MCRT (or high organic loading) and low pH conditions. The excessive growth of fingered zoogloeal colonies may cause settling problems.
- Presence or absence of spirochaetes. These organisms indicate influent wastewater or biological reactor septicity.
- The presence or absence of inorganic and organic particles. These are paper fibers, chemical precipitates, plant ducts, grease particles, and industry-specific particles.
- Neisser-positive cell clumps. They are observed in activated-sludge plants incorporating anoxic or anaerobic selectors or biological phosphorus removal plants. Neisser-positive tetrads occur in plants operated at high MCRTs under nutrient-deficient conditions.
- India ink test. A positive result indicates presence of excessive amounts of extracellular polymers and viscous bulking.
- Chlorine damage to filaments. If RAS chlorination is practiced, excess dosage of chlorine is revealed by empty filament sheaths.
- The presence and effect of filamentous microorganisms on flocs.
    - None.
    - Bridging: the filaments extending from flocs form bridges among them.
    - Open floc structure: the floc-forming organisms attach to filaments, leading to diffuse, irregularly shaped flocs with internal voids.
- Abundance of filamentous microorganisms. Several methodologies exist to quantify the abundance of filamentous organisms. A subjective scoring of filamentous organism abundance is used in the worksheets provided in Figures 8.14 and 8.15. However, specific counting techniques for sludge-bulking and foaming organisms can also be used.
    - Subjective scoring of filamentous organism abundance: filamentous organisms are first observed at 100X and then at 1000X. Their

overall abundance is scored on a scale from 0 (none) to 6 (excessive). A description of the scale is given in Table 8.3 (Jenkins et al., 1993). After the overall abundance of filamentous organisms is determined, the abundance of each filamentous organism is evaluated to determine the types of remedial actions to be taken and the response of each filamentous organism to the remedial action.

— Measurement of TEFL: measuring TEFL (Table 8.4) is tedious and time consuming. Therefore, it is not recommended for monitoring of activated-sludge settleability. On the other hand, because there is a good correlation between SVI and TEFL, SVI can be determined daily to assess variations in activated-sludge settleability. If an increase in SVI is observed, a microscopic examination should be carried out immediately to determine the cause of increase in SVI. If the cause is filamentous organisms, then the causative organisms should be identified to determine the type and extent of the remedial action.

— Measurement of filament count: this method (Table 8.5) was developed by the San Jose/Santa Clara, California, Water Pollution Control Plant staff (Jenkins et al., 1993). It is simple to conduct, but is not necessary for monitoring the settleability of activated sludge.

— Nocardia counting: this method (Table 8.6) is also tedious and time consuming. Therefore, it is not recommended for daily use in

Table 8.3 Subjective scoring of filament abundance (Jenkins et al., 1993).

Numerical value[a]	Abundance	Explanation
0	None	
1	Few	Filaments present, but only observed in an occasional floc.
2	Some	Filaments typically observed, but not present in all flocs.
3	Common	Filaments observed in all flocs, but at low density (e.g., 1–5 filaments per floc).
4	Very common	Filaments observed in all flocs at medium density (e.g., 5–20 per floc).
5	Abundant	Filaments observed in all flocs at high density (e.g., > 20 per floc).
6	Excessive	Filaments present in all flocs—appears more filaments than floc or filaments growing in high abundance in bulk solution.

[a] The scale from 0 to 6 represents to 100 to 1000 fold range of TEFL.

Table 8.4   Measurement of total extended filament length (Sezgin et al., 1978).

Step	Action
1	Transfer 2 mL of a well-mixed activated-sludge sample of known suspended solids concentration using a wide-mouth pipette (0.8-mm diameter tip) to 1 L of distilled water in a 1.5-L beaker stirred at 95 r/min on a jar test apparatus.
2	Mix the content of the beaker for 1 minute.
3	Using the same pipette, transfer 1.0 mL of the diluted sample to a 1 mL Sedgewick Rafter counting chamber with grid lines and cover with a glass cover slip.
4	Using a binocular microscope at 100X with an ocular micrometer scale, count the number of and size filaments present in the whole chamber or a known portion of it and place them in the following size ranges: 0 to 10 μm, 10–25 μm, 25–50 μm, 50–100 μm, 100–200 μm, 200–400 μm, 400–800 μm. Measure filaments of greater than 800 μm in size individually. Count each branch of filaments as a separate filament.
5	Express the results as total filament length in μm per gram MLSS or per milliliter MLSS.

$$\text{TEFL, μm/g MLSS} = \frac{\text{TFL} \times \text{DF}}{\text{MLSS}}$$

or

$$\text{TEFL, μm/ml MLSS} = \text{TFL} \times \text{DF}$$

Where

TFL : total filament length in the 1.0 ml diluted sample, μm; TFL is calculated as the sum of total filament length in each size category and the total filament length in each size category is determined by multiplying the number of filaments in each size category by the average size category length.

DF : dilution factor; 500 (i.e., 1000 mL/2 mL).

MLSS : mixed liquor suspended solids concentration, g/L.

---

assessing the extent of foaming in activated-sludge plants. Instead, the method described for subjective scoring of filamentous organisms abundance can be used.

- The presence and types of protozoa and metazoa. These organisms may be recorded in several categories. For example, protozoa may be divided into five subgroups: amoebae, flagellates, crawling ciliates, free-swimming ciliates, and stalked ciliates. Metazoa include rotifers,

---

Table 8.5 Simplified filament quantification technique (Jenkins et al., 1993).

Step	Action
1	Transfer 50 μL of an activated-sludge sample to a microscope slide.
2	Cover the transferred sample with a 22 × 30 mm cover slip. Care should be taken to spread the sample uniformly under the cover slip.
3	Using a microscope with the eyepiece fitted with a single hairline and starting at the edge of the cover slip, observe consecutive fields at 100X across the entire 30-mm length of the cover slip. (Number of fields may vary with the type of microscope. Some microscopes result in 17 fields).
4	Count the number of times that any filament intersects with the hairline.
5	Sum the number of intersections for all the fields observed across the 30-mm length of the cover slip. This is the filament count for the area of microscope field times 30 mm. If a unit count is desired such as filament count per μL, then the filament count must be multiplied by the number of fields in the 22-mm width of the slide. Assume that there are 12 fields along the 22-mm width of the slide. Thus:

$$\text{Filament count/}\mu L = \frac{\text{number of filaments intersected in the fields along 30-mm length}}{50\mu L}$$

nematodes, and annelids. Identification of these organisms to the species level may not be necessary. However, recognition of these groups is important because each group has a different function and may provide a clue about the condition of the activated sludge. For example, excessive growth of amoebae, flagellates, and some small free-swimming ciliates is associated with dispersed growth (turbid effluent). Turbid effluent may be caused by a variety of conditions, including low MCRT or high F:M, plant startup conditions, presence of toxic substances, low DO, and pH levels outside the range of 6.0 to 8.0. An overabundance of stalked ciliates, rotifers, and other higher forms of life may indicate high MCRT or low F:M operation of the activated-sludge process. If toxicity is present, for example, because of chlorination, heavy metals, or toxic inorganic and organic compounds, the first sign of toxicity may be observed with stalked ciliates and rotifers. The activity of these organisms is first reduced and some may be eliminated. Next, the number of flagellates and small free-swimming ciliates increases. Their increase is caused by floc breakup and the production of a large number of dispersed particles and bacteria. All protozoans and rotifers are finally eliminated if the toxicity is increased.

The abundance of these organisms can be recorded using the methods described in Chapters 5, 6, and 7.

Table 8.6 *Nocardia* filaments counting methodology (Jenkins et al., 1993).

Step	Action
1	Blend 400 mL of mixed liquor for 2 to 3 minutes using low power in a blender.
2	Conduct VSS analysis of the mixed liquor.
3	Mark the edges of 5 to 10 clean frosted microscope slides at three equally spaced points along their length with a pencil.
4	Place 80 μL of the blended mixed liquor using a micropipette on each slide. Spread the sample evenly over the entire nonfrosted area of the slide.
5	Let the slides dry in air.
6	Examine the slides under microscope at 100X using phase contrast to check for even distribution of the mixed liquor solids over the slide. If solids are not distributed evenly on a slide discard it.
7	If less than 5 slides remain, repeat step 1 and steps 3 through 6 to obtain 5 slides having solids distributed evenly.
8	Stain the slides with Gram stain.
9	Count five slides at 1000X using oil immersion under bright field illumination.
10	Counting procedure is as follows: a Locate the pencil mark on the edge of the slide under the microscope at 1000X. b Line up the eyepiece graticule ruled line with the pencil mark on the slide. c Count any intersection with the eyepiece line of Gram-positive branched filaments that are greater than 1 μm in length. d Move across the slide to the opposite edge and count all intersections with the Gram-positive branched filaments greater than 1 μm in length. e Repeat steps a through d at the other pencil marks on the slide. f Take the average of the three counts per slide and express the result as number of intersections/g VSS. g Repeat procedure given in steps a through f for four more slides. h Take the average of all five slides. i The calculation is as follows: *Nocardia* count as number of intersections/g VSS =

$$\frac{\text{Average no. of intersections}}{80 \, \mu L} \times \frac{1000 \, \mu L}{mL} \times \frac{1000 \, mL}{L} \times \frac{1}{MLVSS, g/L}$$

Characterization of Filamentous Organisms. Once the general condition of an activated-sludge sample is evaluated, the next step is to determine the types and abundance of filamentous organisms. Filamentous organisms are characterized on the basis of the several features listed in the worksheet in Figure 8.15. Some of these features are observed under the microscope with oil immersion at 1000X and others at 100X. The features that need to be evaluated are

- Branching: present or absent, and, if present, whether it is true or false branching. True branching means that the cells themselves are forming branches because of the contiguous cytoplasm between branched trichomes (i.e., filaments with or without cells arranged in strands or chains). The filamentous organisms exhibiting true branching in activated sludge are fungi and *Nocardia* spp. *Nostocoida limicola* has been reported to exhibit true branching rarely (Jenkins et al., 1993). False branching refers to trichome branching where the sheath forms branches, not the cell, and therefore the cytoplasm is not contiguous between branched trichomes. *Sphaerotilus natans* is the only organism in activated sludge forming false branches. A closely related organism, type 1701, has also been observed to exhibit false branching in pure cultures only (Jenkins et al., 1993).
- Motility: none or present. Several filamentous organisms present in activated sludge exhibit motility. *Beggiatoa* spp., *Flexibacter* spp., *Herpetosiphon* spp., and some blue-green bacteria (*Cyanophyceae*) move by gliding. *Thiothrix* spp. and some type 021N may exhibit limited twitching or swaying motions (Jenkins et al., 1993). *Microthrix parvicella* cells or short filaments in pure cultures have been observed to be very motile (Wanner, 1994).
- Filament shape: several terms are used to describe the shape of filamentous organisms: straight, smoothly curved, bent, chain of cells with an irregular shape, coiled, and mycelial (for example, fungi and *Nocardia* spp.).
- Location: filaments in activated sludge may be observed in three locations: extending from the floc surface, mostly within the floc, and free-floating in the bulk liquid between flocs.
- Attached growth: some filaments are covered with unicellular bacteria. This feature is used in differentiation of certain filamentous organisms. Record the result of the microscopic examination as whether attached growth is present or absent. If present, determine whether the growth is substantial or incidental. Some filaments may exhibit occasional attached growth, such as blue-green bacteria (Cyanophyceae) but this feature is not used in their identification.
- Sheath: present or absent. The presence of a sheath is used in identification of several filamentous organisms. However, it is difficult to see because it is a clear structure (exterior to the cell wall). There are several ways to establish its presence or absence. They can be observed in wet mounts without any staining when trichomes are empty (devoid of cells) or have empty spaces between cells. In this latter case, the

---

sheath can be observed continuing along either side of the empty space. There are also several features of filamentous organisms that can be confused with a sheath. For example, when a sample is examined under phase contrast, the sheath can be mistaken for a yellowish halo observed around filaments that actually results from light reflection. Short, empty spaces in a trichome or at the apex should not be confused with sheaths. The cell wall after cell lysis also should not be confused with sheaths. However, a cell wall can be distinguished from a sheath by observing the presence of a cell septum. Presence of attached growth typically indicates presence of a sheath.

There are several ways to detect presence or absence of a sheath. An activated-sludge sample is mixed with a diluted (1:1000 dilution) household bleach (sodium hypochlorite solution) in equal volumes. The mixture is then allowed to stand for several hours or overnight. This treatment causes cell lysis inside the sheath, resulting in empty sheaths that can be observed under phase contrast at 1000X (Jenkins et al., 1993). Similarly, a sample of activated sludge is aerated for approximately 2 to 3 weeks without feeding. This treatment also results in cell lysis and empty sheaths can be observed under phase contrast at 1000X. Staining of an activated-sludge sample with crystal violet (Figure 8.12)(Jenkins et al., 1993) or India ink (Figure 8.10)(Wanner, 1994) may facilitate the detection of a sheath.

- Cross-walls (cell septa): present or absent. This feature may be variable for some filamentous organisms. Detection of cell walls requires a high-resolution microscope. It is important to determine whether a filamentous organism is made up of a chain of cells or a true trichome.
- Filament diameter: record the average diameter and also its range in micrometers.
- Filament length: record the range in micrometers.
- Cell shape: the shape of the cells can be square, rectangular, oval, barrel, discoid, round-ended rods, or irregular. The other important features of the cells used in identification are whether there are indentations at cell septa or whether trichome walls are straight at the cell junctions.
- Cell size: report the average and range of the length and width of the cells in micrometers.
- Sulfur deposits: if present, sulfur granules appear as bright, refractive yellow-colored cell inclusions under phase-contrast illumination. Report sulfur granules as present or absent in situ and present or absent after carrying out the sulfur oxidation test. The shape of the granules provides a clue about the identity of certain filamentous organisms. Report also the shape of sulfur granules, whether spherical or square and rectangular. The sulfur granules are spherical in *Thiothrix* spp., *Beggiatoa* spp., and type 021N and are square or rectangular in type 0914. Type 0914 does not respond to the sulfur oxidation test.
- Other intracellular granules: report as present or absent. These are polyphosphate and PHB granules. Polyphosphate granules can be

detected by Neisser staining (Figure 8.8) and PHB granules by PHB staining (Figure 8.11).

- Staining reactions: carry out Gram and Neisser staining as described in Figures 8.7 and 8.8, respectively. Note the general location, diameter, length, cell shapes, and presence or absence of attached growth in the wet mount so that the same filament types can be examined in the stained smears. Some filamentous organisms change their size upon staining. For example, type 0092 appears wider in Neisser-stained smears than in wet mounts (Jenkins et al., 1993). Therefore, care should be taken not to confuse these organisms with others. Type 0041 and *Thiothrix* spp., may stain Neisser positive in papermill activated-sludge plants.

  Record the results of the Gram staining as strongly positive, weakly positive, variable, or negative. When conducting a Gram staining, include fresh cultures with known Gram-staining reactions as positive or negative controls. Most filamentous organisms in activated sludge are Gram negative. Filamentous organisms, *M. parvicella* and *Nocardia* spp., are typically strongly Gram positive. When some filaments (i.e., type 0041 and type 0675) are covered with a substantial attached growth, the Gram staining of the trichome may be difficult to observe or it may appear to be Gram negative because of a Gram-negative reaction of the attached growth. *Nostocoida limicola* and types 0041 and 0675 are typically Gram positive but can be Gram variable or Gram negative (Jenkins et al., 1993). *Thiothrix* I, *Beggiatoa* spp., type 021N, and type 0914 typically are Gram negative but may stain Gram positive when they contain a substantial amount of intracelluler sulfur granules (Jenkins et al., 1993).

  Neisser staining is recorded as negative when the entire trichome, including the intracellular granules, is stained as yellow-brown and positive when the entire trichome is purple or blue-violet. Variations in the intensity of the stain are sometimes observed. For example, type 0092 is stained light purple and *N. limicola* dark purple (Jenkins et al., 1993). The trichome is sometimes stained as yellow-brown and intracellular granules as blue-violet, as in the case with *M. parvicella* and *Nocardia* spp. In this case, the organism is reported as having Neisser-negative trichome with Neisser-positive intracellular granules. Filamentous organisms, *Beggiatoa* spp., *Thiothrix* spp., and types 0041, 0675, 021N, 0914, and 1863 may occasionally exhibit Neisser-positive intracellular granules (Jenkins et al., 1993). *Haliscomenobactor hydrosis* and types 0041 and 0675 may have a Neisser-positive trichome covering in activated-sludge plants treating industrial wastewaters (Jenkins et al., 1993).

- Other common observations: *Thiothrix* spp. and (less often) type 021N exhibit gonidia and rosette formations. Gonidia form under environmental conditions unfavorable to rapid growth. The individual cells of the trichome become rod-shaped or oval toward the tip of the filament and are ultimately released to the environment as individual cells. A rosetta

---

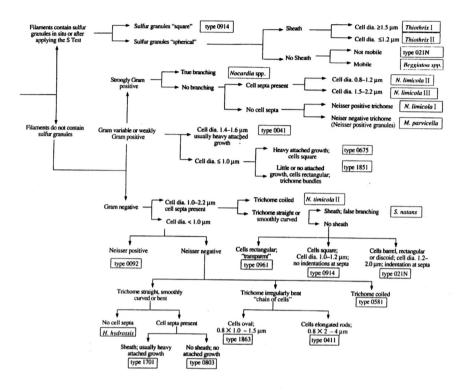

Figure 8.16   Identification key for filamentous organisms present in activated sludge (Jenkins et al., 1993).

formation occurs when filaments radiate outward from a common origin. Gonidia and rosettes are formed in nutrient-deficient conditions and also observed in plants treating septic wastewater (Jenkins et al., 1993).

Identification of Filamentous Organisms. Once the characteristics of the filamentous organisms observed in an activated-sludge sample are recorded on the worksheet provided in Figure 8.15, the organisms can be identified using the key given in Figure 8.16 (Jenkins et al., 1993). In addition, Table 8.7 (Jenkins et al., 1993), which summarizes the morphological and staining characteristics of filamentous organisms, can be used to aid in identification of organisms. Twenty-two organisms are listed in the table and also described in the key. The filamentous organisms types 1702, 1852, and 0211 are observed occasionally, and therefore omitted from the key and table. They are described briefly under the next section. Several other filamentous organisms that have easily identifiable features are omitted from the key to further simplify it but also are described in the next section. These organisms include fungi, *Cyanophyceae*, *Flexibacter* spp., and *Bacillus* spp. Some filamentous organisms were entered in the key twice because of their variable characteristics. For example, *N. limicola* II exhibit both positive and negative reactions to Gram staining. Types 0914 and 021N may contain sulfur granules in situ.

When sulfur granules are absent from in situ samples, type 0914 does not respond to the sulfur test and the response of 021N is variable, but mostly positive (Jenkins et al., 1993).

Table 8.8 provides a summary of the morphological and staining characteristics of filamentous organisms described by Eikelboom and van Buijsen (1981). There are certain differences in the description of the same organisms given in Tables 8.7 and 8.8. These differences include variations in filament dimensions, cell shape and size, cell inclusions, and Gram and Neisser staining.

The characteristics of a filamentous organism recorded on the worksheet (Figure 8.15) should be checked against the characteristics of organisms provided in Figure 8.16 and Table 8.7, respectively, and the next section. If the filamentous organism still cannot be identified, certain characteristics of the filamentous organism such as Gram and Neisser staining should be reexamined. Table 8.8 should also be consulted because it presents further variations in morphology and staining characteristics of certain filamentous organisms. If a filamentous organism cannot be identified, it should be listed as not identified and not be forced into a type or genus.

A few new filamentous organisms have been described in the literature (Scheff et al., 1984; Senghas and Lingens, 1985; Trick and Lingens, 1984; and Trick et al., 1984) since publication of the filament identification keys. These filamentous organisms are also described in the next section.

Description of Filamentous Organisms. A short description of filamentous organisms based on information given in Jenkins et al. (1993) and Eikelboom and van Buijsen (1981) are presented below.

SPHAEROTILUS NATANS. Filaments are straight or smoothly curved and composed of round-ended, rod-shaped cells within a sheath. The cells are rectangular when they are tightly fitted in the sheath. The cell size varies from 1.0 to 1.8 × 1.5 to 3.0 µm according to Jenkins et al. (1993) or 1.2 to 2.0 × 1.5 to 5.0 µm according to Eikelboom and van Buijsen (1981). Cell septa are observed with indentations at the septa. The filaments exhibit false branching, which gives the appearance of tree formation. They radiate from the floc surface. They are Gram negative, Neisser negative, frequently contain spherical PHB granules (typically three per cell), and do not contain sulfur granules. Filament length is greater than 500 µm. Attached growth is frequently absent but may be observed when the organism is growing slowly.

TYPE 1701. Filaments are relatively short (20 to 200 µm), curved or bent filaments containing round-ended, rod-shaped cells with dimensions of 0.7 to 1.0 to 1.0 to 2.0 µm (Jenkins et al., 1993) or 0.7 to 0.9 × 2.5 to 3.5 µm (Eikelboom and van Buijsen, 1981). The cells are surrounded with a sheath. Filaments are typically observed within flocs, but a few extend into the bulk medium from the floc surface. Attached growth is often observed but may be absent when filaments are growing fast. False branching may occur but is

Table 8.7 Summary of typical morphological and staining characteristics of filamentous organisms observed in activated sludge (Jenkins et al., 1993).

Filament Type	Gram stain	Neisser Trichome	Neisser Granules	Sulfur granules in situ	Sulfur granules S test	Other cell inclusions	Trichome diameter, μm	Trichome length, μm	Trichome shape	Trichome location	Cell septa clearly observed	Indentations at cell septa	Sheath	Attached growth	Cell shape and size, μm	Notes
S. natans	-	-	-	-	-	PHB	1.0–1.4	>500	St	E	+	+	+	-	Round-ended rods 1.4 × 2.0	False branching
type 1701	-	-	-	-	-	PHB	0.6–0.8	20–80	St, B	I, E	+	+	+	++	Round-ended rods 0.8 × 1.2	Cell septa hard to discern
type 0041	+, V	-	+	-	-	-	1.4–1.6	100–500	St, SC	I, E	+	-	+	++, -	Squares 1.4 × 5–2.0	Neisser-positive reaction occurs
type 0675	+, V	-	+	-	-	-	0.8–1.0	50–150	St, SC	I	+	-	+	++, -	Squares 1.0 × 1.0	Neisser-positive reaction occurs
type 021N	-	-	+	-, +	-, ++	PHB	1.0–2.0	50–1000	St, SC	E	+	+	-	-	Barrels, rectangles discoid 1.0–2.0 × 1.5–2.0	Rosettes, gonidia
Thiothrix I	-, +	-	+	+, -	+	PHB	1.4–2.5	100->500	St, SC	E	+	-	+	-	Rectangles 2.0 × 3.0–5.0	Rosettes, gonidia
Thiothrix II	-	-	+	+, -	+	PHB	0.8–1.4	50–200	St, SC	E	+	-	+	-	Rectangles 1.0 × 1.5	Rosettes, gonidia
type 0914	-, +	-	+	-, +	-	PHB	1	50–200	St	E, F	+	-	-	-	Squares 1.0 × 1.0	Sulfur granules square
Beggiatoa sp.	-, +	-	+	+, -	+	PHB	1.2–3.0	100->500	St	F	-, +	-	-	-	Rectangles 2.0 × 6.0	Motile; flexing and gliding
type 1851	+, weak	-	+	-	-	-	0.8	100–300	St, SC	E	+, -	-	+	-, +	Rectangles 0.8 × 1.5	Trichome bundles
type 0803	-	-	-	-	-	-	0.8	50–150	St	E, F	+	-	-	-	Rectangles 0.8 × 1.5	

Direct illumination observation

Phase-contrast observation at 1000 X

Table 8.7 Summary of typical morphological and staining characteristics of filamentous organisms observed in activated sludge (Jenkins et al., 1993) (continued).

Filament Type	Gram stain	Neisser stain Trichome	Neisser stain Granules	Sulfur granules in situ	Sulfur granules S test	Other cell inclusions	Trichome diameter, µm	Trichome length, µm	Trichome shape	Trichome location	Cell septa clearly observed	Indentations at cell septa	Sheath	Attached growth	Cell shape and size, µm	Notes
type 0092	–	+	–	–	–	+	0.8–1.0	20–60	St, B	I	+, –	–	–	–	Rectangles 0.8 × 1.5	
type 0961	–	–	–	–	–	–	0.8–1.2	40–80	St	E	+	–	–	–	Rectangles 0.8–1.4 × 2.0–4.0	Transparent
M. parvicella	+	–	+	–	–	PHB	0.8	50–200	C	I	–	–	–	–	–	Large patches
Nocardia sp.	+	–	+	–	–	PHB	1	30-May	I	I	+, –	–	–	–	Variable 1.0 × 1.0–2.0	True branching
N. limicola I	+	+	–	–	–	–	0.8	100	C	I, E	–	–	–	–	Discs, ovals 1.2 × 1.0	
N. limicola II	–, +	+, –	–	–	–	PHB	1.2–1.4	100–200	C	I, E	+	+	–	–		Incidental branching, Gram & Neisser variable
N. limicola III	+	+	–	–	–	PHB	2	200–300	C	I, E	+	+	–	–	Discs, ovals 2.0 × 1.5	
H. hydrossis	–	–	–	–	–	–	0.5	10–100	St, B	E, F	–	–	+	–, +	–	Rigidly straight
type 0581	–	–	–	–	–	–	0.5–0.8	100–200	C	I	–	–	–	–		
type 1863	–	–	–, +	–	–	–	0.8	20–50	B, I	E, F	+	+	–	–	Oval rods 0.8 × 1.0–1.5	Chain of cells
type 0411	–	–	–	–	–	–	0.8	50–150	B, I	E	+	+	–	–	Elongated rods 0.8 × 2.0–4.0	Chain of cells

LEGEND: + = positive; – = negative; V = variable; +, – or –, + = variable, the first being most common; single symbol = invariant; Trichome Shape: St = Straight; B = Bent; SC = Smoothly curved; C = Coiled; I = Irregularly shaped; Trichome Location: E = Extending from the floc surface; I = Inside the floc; F = Free in liquid between the flocs

Table 8.8 Summary of typical morphological and staining characteristics of filamentous organisms observed in activated sludge (adapted from Eikelboom and van Buijsen, 1981).

| | Direct illumination observation | | | | | | Phase-contrast observation at 1000 X | | | | | | | | | |
Filament Type	Gram stain	Neisser stain Trichome	Neisser stain Granules	Sulfur granules in situ	Sulfur granules S test	Other cell inclusions	Trichome diameter, μm	Trichome length, μm	Trichome shape	Trichome location	Cell septa clearly observed	Indentations at cell septa	Sheath	Attached growth	Cell shape and size, μm	Notes
S. natans	−	−	−	−	−	−, + PHB	1.2–2.0	500–1000	St, SB	E	+	+	+	−, +	Rods and rectangles 1.2–2.0 × 1.5–5.0	False branching
type 1701	−	−	−	−	−	−, + PHB	0.7–0.9	100–200	St, B	I, E	+	+	+	+	Rods 0.7–0.9 × 2.5–3.5	False branching occurs incidentally
type 0041	+, V	−		−	−, +	−	1.0–1.4	200–300	St, SB	E, F	+	−	+	−, +	Square to rectangles 1.0–1.4 × 0.7–2.3	Incidental branching occurs
type 0675	+, V	−		−	−, +	−	0.5–0.7		St, SB	E	+	−	+	−, +	Square to rectangles	
type 021N	−	−		−	−, +	−	0.6–0.8 1.8–2.2	500–1000	St, SB	E	+	+	−	−	Discs, 1.8–2.2 × 0.4–0.7 Rods, 0.6–0.8 × 2.0–3.0 mostly square	Intermediate cell sizes are possible. Forms rosettes.
Thiothrix I	−	−		+, −	+	−	0.4–1.5	50–500	St, SB	E	+	−	−	−	Rectangles	Rosettes, gonidia
Thiothrix II	−	−		+, −	+	−	0.8–1.5	200–800	St, SB	E	+	−	+	−	Rectangles	Rosettes, gonidia
type 0914	+	−		−, +	−	−	0.7–0.9	100–200	St, SB	F	+	−	−	−	Squares and rectangles	
Beggiatoa sp.	−	−		+	+	−	<1.0–>2.5	<200	St, SB	F	−, +	−	−	−	Square or rectangle	Motile: flexing and gliding
type 1851	+, weak	−		−	−	−	0.5–0.7	200–400	St, SB		+, −	−	+	+	Rectangles 0.5–0.7 × 1.7–3.5	Trichome bundles
type 0803	−	−		−	−	−	0.7–0.8	100–300	St, B	F, E	−, +	−	−	−	Squares to rectangles	

Table 8.8 Summary of typical morphological and staining characteristics of filamentous organisms observed in activated sludge (adapted from Eikelboom and van Buijsen, 1981) (continued).

	Direct illumination observation						Phase-contrast observation at 1000 X									
Filament Type	Gram stain	Neisser stain Trichome	Neisser stain Granules	Sulfur granules in situ	Sulfur granules S test	Other cell inclusions	Trichome diameter, μm	Trichome length, μm	Trichome shape	Trichome location	Cell septa clearly observed	Indentations at cell septa	Sheath	Attached growth	Cell shape and size, μm	Notes
type 0092	–	+	–	–	–	–	0.5–0.7	<100	St, B	I, F	–, +	–	–	–	Rectangles	
type 0961	–	–	–	–	–	–	1.1–1.5	300–500	St	E	+	–	–	–	Rectangles	Transparent
M. parvicella	+	–	–	–	–	+P	0.5	200–400	B	I	–	–	–	–		
Nocardia sp.	+	–	+	–	–	–	0.5	<100	I	I	–	–	–	–		True branching
N. limicola I	+	+	–	–	–	–	0.6–0.7	100–300	C	F	–, +	–	–	–	Spherical cells	
N. limicola II	+	+	–	–	–	–	1.0–1.2	100–300	C	F	+	+	–	–	Spheres to discs	
N. limicola III	+	+	–	–	–	–	1.5		C	F	+	+	–	–	Disc-shaped	
H. hydrossis	–	–	–	–	–	–	0.3	<100	St	E	–	–	+	–, +		Needle like
type 0581	–	–	–	–	–	–	0.3–0.4	100–300	B, T	F	–	–	–	–		Chain of cells
type 1863	–	–	–	–	–	–	0.8	<150	B	F	+	+	–	–	Cocus, rod-shaped	Chain of cells
type 0411	–	–	–	–	–	–	0.5–0.7	50–150	B, T	E	+	+	–	–	Long rod-shaped cells	Chain of cells

LEGEND: + = positive; – = negative; V = variable; +, – or –, + = variable, the first being most common; single symbol = invariant; Trichome Shape: St = Straight; SB = Smoothly bent; B = Bent; I = Irregulary shaped; C = Coiled; T = Twisted; Trichome Location: E = Extending from the floc surface; I = Inside the floc; F = Free in liquid between the flocs rare. They are Gram negative, Neisser negative, have no sulfur granules, and frequently contain spherical PHB granules.

TYPE 0041. Filaments are straight or smoothly curved, 100 to 500 μm in length, and have sheaths. Cells are 1.2 to 1.6 × 1.5 to 2.5 μm in dimension (Jenkins et al., 1993) and 1.0 to 1.4 to 0.7 × 2.3 μm (Eikelboom and van Buijsen, 1981). Cells are square to rectangular in shape and exhibit no indentations at cell septa. A larger form of type 0041, with a trichome diameter of 2.5 to 4.0 μm, is sometimes observed (Jenkins et al., 1993). Heavy attached growth is often observed. In industrial wastewater activated-sludge plants, type 0041 may occur extending from the floc surface or in the bulk solution and may not have any attached growth (Jenkins et al., 1993). Type 0041 is Gram positive and may sometimes exhibit variable Gram staining. It tends to be Gram positive when found within the floc and negative when extending from the floc surface into bulk solution. Neisser staining is negative, but Neisser-positive granules are sometimes observed. When a slime coating is present around the trichomes, it has a Neisser-positive layer around the trichome. Intracellular granules are seldom observed. No sulfur granules are present. However, sulfur granules may be observed after the sulfur test (Eikelboom and van Buijsen, 1981). The sheath is difficult to observe with an ordinary light microscope and is best to see when some cells are missing. The trichome of type 0041 resembles the trichomes of colorless *Cyanophyceae* or type 021N when type 0041 does not contain attached growth. However, the differences between type 0041 and *Cyanophyceae* are that *Cyanophyceae* are mostly motile and do not have a sheath and type 0041 is not motile and has a sheath. On the other hand, the presence of a sheath is difficult to detect and not a reliable property to distinguish between the two organisms. Type 0041 can be differentiated from type 021N on the basis of cell shape, Gram staining, indentations at cell septa, rosette formation, and the presence of a sheath. The shapes of the cells of type 0041 are square to rectangle and those of type 021N are barrels, discoids, and rectangles. Type 021N is Gram negative and type 0041 is either Gram positive or Gram variable. Type 021N does not have a sheath. Type 0041 may also resemble *Thiothrix* I in industrial WWTPs.

TYPE 0675. Type 0675 is typically observed together with type 0041. It is similar to type 0041 morphologically and in staining characteristics, but has smaller trichome diameter and length. Cells are 0.8 to 1.0 μm in diameter and 1.0 mm in length (Jenkins et al., 1993). Cell diameter is 0.5 to 0.7 μm (Eikelboom and van Buijsen, 1981). Trichome length ranges from 50 to 150 μm (Jenkins et al., 1993). Trichomes are often covered with heavy attached growth, but attached growth may be lacking on them when observed in some industrial wastewater activated-sludge plants. They have a sheath and are Gram positive but may exhibit Gram-variable results. They are Neisser negative and Neisser-positive granules may occur (Jenkins et al., 1993). Sulfur granules are absent (Jenkins et al., 1993) but may be observed after the sulfur test (Eikelboom and van Buijsen, 1981).

TYPE 021N. Trichomes are straight, smoothly curved or slightly coiled (Jenkins et al., 1993) or smoothly bent (Eikelboom, 1975). Cells are 1.0 to 2.0 μm in diameter and 1.5 to 2.0 μm in length and barrel, discoid, or rectangular in shape (Jenkins et al., 1993). According to Eikelboom and van Buijsen (1981), when cells are in disc shapes, their sizes range from 1.8 to 2.2 × 0.4 to 0.7 μm and, when they are rods, their dimensions are 0.6 to 0.8 × 2.0 to 3.0 μm. Intermediate sizes are also observed. Cells have indentations at cell septa. Trichome length ranges between 100 to greater than 1000 μm. The trichome sometimes tapers from a thicker basal region to a thinner apical region, which terminates with an attached gonidia. Trichomes occasionally may have a knot. They are Gram negative and Neisser negative and may contain Neisser-positive granules. They may stain slightly Gram positive in the presence of sulfur granules. Spherical sulfur granules are occasionally observed in in situ samples and their response to the sulfur test is typically positive. However, according to Eikelboom and van Buijsen (1981), type 021N does not contain sulfur granules in in situ samples. Trichomes form rosettes infrequently and do not contain attached growth and have no sheath.

THIOTHRIX I. Trichomes are straight or smoothly curved and 100 to 500 μm in length. Filaments are observed extending from the floc surface. Trichome diameter ranges between 1.4 to 2.5 μm according to Jenkins et al. (1993) and 0.4 to 1.5 μm according to Eikelboom and van Buijsen (1981). Cells are rectangular in shape with dimensions of 1.4 to 2.5 × 3 to 5 μm with clearly observable crosswalls without indentations at cell septa (Jenkins et al., 1993). Although it has been reported (Jenkins et al., 1993, and Wanner, 1994) that this organism has a heavy sheath, Eikelboom and van Buijsen (1981) indicate that this organism does not have any sheath. They are Gram and Neisser negative but may be Gram positive in the presence of sulfur granules. Neisser-positive granules may be observed. They may stain Neisser positive when growing in papermill WWTPs. Spherical sulfur granules may be present in situ samples. They respond to the sulfur test strongly. Rosettes and gonidia may be observed.

THIOTHRIX II. Filaments are straight or smoothly curved, 50 to 200 μm (Jenkins et al., 1993) or 200 to 800 μm in length (Eikelboom and van Buijsen, 1981). Trichomes have sheaths and typically no attached growth. A small amount of attached growth may be observed when the filament is not growing. Cells are rectangular in shape and have no indentations at cell septa. Cell sizes range from 0.7 to 1.4 × 1 to 2 μm. They are Gram and Neisser negative with occasional Neisser-positive granules. Cells typically contain spherical sulfur granules in situ and result in positive reaction to the sulfur test. Filaments frequently have rosettes and gonidia.

TYPE 0914. Filaments are straight or smoothly curved and 50 to 200 μm in length, found both extending from the floc surface and free in the bulk solution (Jenkins et al., 1993). Cells are square and rectangular in shape with dimensions of 0.7 to 0.9 × 1.0 μm and have no indentations at cell septa. No sheath is present and some attached growth may be observed incidentally. They are Gram negative but may stain Gram positive in the presence of sulfur granules (Jenkins et al., 1993). These organisms have been described as Gram positive by Eikelboom and van Buijsen (1981). They are Neisser negative and Neisser-positive granules may occur. Sulfur granules may be present in situ and, if present, they are square in shape. They do not respond to the sulfur test.

BEGGIATOA SPP. Filaments are straight, 100 to 500 μm in length, and observed in bulk solution gliding and flexing. Cells are rectangular in shape with sizes of 1 to 3 × 4 to 8 μm (Jenkins et al., 1993). The cell length may be as large as 10 to 20 μm. Filaments may contain significant amounts of sulfur granules. It is difficult to observe the cell septa in the presence of sulfur granules. These organisms respond to the sulfur test. Filaments have no sheath and no attached growth. They are Gram negative but may stain Gram positive in the presence of sulfur granules. They are Neisser negative and may have Neisser-positive granules.

TYPE 1851. Trichomes are straight or smoothly curved and observed extending from the floc surface, typically in bundles of intertwined filaments. Cell sizes range between 0.8 to 1.0 × 1.5 to 2.5 μm according to Jenkins et al. (1993) and 0.5 to 0.7 × 1.7 to 3.5 μm according to Eikelboom and van Buijsen (1981). Cells are rectangular in shape. Filament length ranges between 100 and 400 μm. The organisms have a sheath and typically have sparse attached growth. Filaments are either weakly Gram positive or Gram negative and Neisser negative. No sulfur granules are observed.

TYPE 0803. Filaments are straight or smoothly curved and 50 to 300 μm in length. They are found either extending from the floc surface or, less often, in the bulk solution. Cells are rectangular (0.8 × 1.5 to 2.0 μm) in shape without any indentations at cell septa (Jenkins et al., 1993). Neither a sheath nor attached growth is present. They are Gram and Neisser negative and no sulfur granules are observed.

TYPE 0092. Filaments are straight, irregularly curved, or bent and found mostly within flocs and occasionally in the bulk solution. Filaments are less than 100 μm in length (Eikelboom, 1975). Cell diameter ranges between 0.8 and 1.0 μm according to Jenkins et al. (1993) and 0.5 and 0.7 μm according to Eikelboom and van Buijsen (1981). Cells are rectangular (0.8 to 1.0 × 1.5 μm) in shape without indentations at cell septa (Jenkins et al., 1993), although cell septa are sometimes hard to observe. No attached growth or sheath is present. They are Gram negative and Neisser positive; no sulfur granules are observed. Filaments appear wider (in the range of 1.0 to 1.2 μm) in dried, Neisser- and Gram-stained smears than when observed in wet mounts (Jenkins et al., 1993).

TYPE 0961. Trichomes are straight and observed extending from the floc surface. Filament length ranges between 20 and 150 μm (Jenkins et al., 1993) or 300 and 500 μm (Eikelboom and van Buijsen, 1981). Cell diameter ranges between 0.8 and 1.4 μm (Jenkins et al., 1993) or 1.1 and 1.5 μm (Eikelboom and van Buijsen, 1981). Cells are rectangular (0.8 to 1.4 × 2 to 4 μm) in shape without indentations at cell septa. Neither attached growth nor a sheath is present, but a slime coating may be observed. The cells appear to be transparent without any internal structures. They are Gram and Neisser negative and no sulfur granules occur.

MICROTHRIX PARVICELLA. Filaments are irregularly coiled or bent and observed within the floc or as patches free in the bulk solution. The trichome diameter is 0.5 μm according to Eikelboom and van Buijsen (1981) and ranges between 0.6 and 0.8 μm according to Jenkins et al. (1993). Different ranges of trichome length have been reported by investigators. Jenkins et al. (1993) have reported a trichome length in the range of 50 to 200 μm, and Eikelboom and van Buijsen (1981) reported length of 200 to 400 μm. Cell septa are not observable. Attached growth and a sheath are absent. However, Eikelboom and van Buijsen (1981) have reported the occasional presence of attached growth. Intracellular granules are observed. They are strongly Gram positive and Neisser negative but have Neisser-positive granules.

NOCARDIA SPP. Filaments are irregularly bent and short (5 to 30 μm) in length, found mostly within the floc, but also free in the bulk solution. They exhibit true branching. Trichome diameter ranges between 0.5 and 1.0 μm. Cell septa are sometimes observable. Neither attached growth nor a sheath is found. They stain Gram positive and Neisser negative and contain Neisser-positive granules. Also, PHB granules are frequently observed, but no sulfur granules are present. Some common species are *Nocardia amarae* and *Nocardia pinensis. Nocardia pinensis* exhibits pine-tree-like branching.

NOSTOCOIDA LIMICOLA I. Filaments are bent and irregularly coiled, 100 to 300 μm in length, and 0.6 to 0.8 μm in diameter, found mostly within the floc and free in the bulk solution. Cell septa are hard to observe. Cells, when visible, are oval and spherical. No sheath, attached growth, or sulfur granules are found. They are Gram and Neisser positive.

NOSTOCOIDA LIMICOLA II. Filaments are bent or irregularly coiled, 100 to 300 μm in length, found within the floc and free in the bulk solution. Cell septa are clearly observable. Cells are spherical, oval, and discoid in shape. Trichome diameter ranges between 1.0 and 1.2 μm according to Eikelboom and van Buijsen (1981) and 1.2 and 1.4 μm according to Jenkins et al. (1993). Cell dimensions are 1.2 × 1.0 to 1.2 μm. No sheath, attached growth, or sulfur granules are present. Also, PHB granules are generally observed. They are typically Gram negative but sometimes Gram positive. Eikelboom and van Buijsen (1981) have reported this organism as Gram positive. The trichome is Neisser positive but may occasionally stain Neisser negative. Jenkins et al. (1993) have reported another form of *N. limicola* II found in some industrial wastewater activated-sludge plants, with Gram-negative and Neisser-negative staining characteristics.

NOSTOCOIDA LIMICOLA III. Filaments are bent and irregularly coiled, 200 to 300 μm in length, and typically found extending from the floc surface. Cells are spherical, oval, and discoid in shape and 1.5 to 2.0 μm in diameter. *Nostocoida limicola* III filaments resemble strings of pearls. No attached growth, sheath, or sulfur granules occur, but PHB granules are observed. They are Gram positive and Neisser positive, but Gram-negative and Neisser-negative reactions may occasionally be observed.

HALISCOMENOBACTER HYDROSSIS. Filaments are straight and bent, 0.3 to 0.5 μm in diameter, and 10 to 100 μm in length, found radiating from the floc surface. They have a needlelike and pinlike appearance. Cell septa are not observed. A sheath is present. Attached growth is typically observed. No sulfur granules occur. Their responses to Gram and Neisser staining are negative.

TYPE 0581. Filaments are smoothly coiled and bent, 100 to 300 μm in length, and found mostly within the floc but may be observed as free patches in bulk solution. The trichome diameter ranges between 0.3 and 0.4 μm according to Eikelboom and van Buijsen (1981) and 0.4 and 0.7 μm according to Jenkins et al. (1993). No sheath, cell septa, attached growth, or sulfur granules occur. Their responses to Gram and Neisser staining are both negative. This organism is similar to *M. parvicella*, but Gram and Neisser staining are different than those of *M. parvicella*.

TYPE 1863. Filaments are irregularly bent or coiled, less than 150 μm in length, and found extending from the floc surface and in the bulk solution. Cells are oval-shaped rods or cocci (0.8 × 1.5 μm) and form chains of cells. No sulfur granules are observed. They are Gram negative and Neisser negative. Neisser-positive granules may be observed.

TYPE 0411. Trichomes are irregularly bent, 0.5 to 0.8 μm in diameter and 50 to 150 μm in length, and observed extending from the floc surface. Cells are elongated rod shapes, with dimensions of 0.5 to 0.8 × 2 to 4 μm. No sheath or attached growth is observed. No sulfur granules are present. They are Gram and Neisser negative.

TYPE 1702. Filaments are straight or bent, 0.6 to 0.7 μm in diameter and 20 to 80 μm in length, and observed within the floc or extending from the floc. Cell septa are not visible. No attached growth, and no sulfur granules occur. A sheath is present. They are Gram and Neisser negative.

TYPE 1852. Trichomes are straight, slightly bent, 0.6 to 0.8 μm in diameter, 20 to 80 μm in length, and observed extending from the floc surface. Cells are rectangular in shape with a dimension of 0.6 to 0.8 × 1.0 to 2.0 μm. No sheath, attached growth, or sulfur granules are found. Cells appear to be transparent. They are Gram negative and Neisser negative.

TYPE 0211. Filaments are bent and twisted, 0.2 to 0.5 µm in diameter, 20 to 100 µm in length, and extend from the floc surface or are free in the bulk solution. The cells are rod shaped with indentations at cell septa. No attached growth, sheath, or sulfur granules occur. The organism is both Gram and Neisser negative.

FLEXIBACTER SPP. Filaments are straight or smoothly curved, 1.0 µm in diameter, 20 to 40 µm in length, and found free in bulk solution. The filaments are motile by gliding and flexing. Cell septa are not visible. No sheath, attached growth, or sulfur granules exist. Typically, PHB granules are observed. The organism is both Gram and Neisser negative.

BACILLUS SPP. Filaments consist of rod-shaped chains of cells, 0.8 to 1.0 µm in diameter, 20 to 50 µm in length, and found mostly at the edges of flocs. No sheath or sulfur granules are present. They are both Gram and Neisser negative.

CYANOPHYCEAE. Filaments are straight, 2 to 5 µm in diameter, 100 to 1000 µm in length, and found free in bulk solution. The organism is motile by gliding. Cells are square to rectangular (2 to 5 × 2 to 8 µm) in shape. Cell septa are clearly visible. Trichomes have no indentations at cell septa. No sheath, attached growth, or sulfur granules are observed. The Gram and Neisser stains are both negative. However, slightly Gram-positive staining is occasionally observed.

FUNGI. These organisms have irregularly shaped filaments, 2 to 8 µm in diameter, 200 to 1000 µm in length, and are found mostly in the floc. Filaments exhibit true branching. Cells are rectangular in shape with dimensions of 2 to 8 × 5 to 15 µm. The Gram and Neisser stainings of the organism are both negative. No sulfur granules and no sheath occur.

HERPETOSIPHON SPP. Trichomes are straight, smoothly curved, bent, 0.5 to 1.5 µm in diameter, and 200 to 500 µm in length. They have a sheath but no attached growth. They are motile by gliding. Cells are rod shaped or rectangular (0.5 to 1.5 × 3 to 5 µm). They have PHB granules but no sulfur granules. The Gram stain is negative.

TRICHOCOCCUS SPP. Filaments are coiled and form chains of cells. Cells are spherical to ovoid (1.0 to 1.5 × 1.0 to 2.5 µm) in shape. Cell septa are clearly visible. The Gram stain is positive.

STREPTOCOCCUS SPP. Filaments are coiled, 0.7 to 0.8 µm in diameter, and less than 100 µm in length. They are found within the floc. Cells are coccus shaped with indentations at cell septa. Filaments are formed from a chain of cells. No attached growth and no sulfur granules are observed. The Gram stain is positive and the Neisser stain is negative.

BACILLUS MYCOIDES AND BACILLUS CEREUS. These organisms have been isolated from activated sludge (Trick et al., 1984), but their staining and morphological characteristics are described after the growth on artificial medium.

FACTORS AFFECTING GROWTH OF FILAMENTOUS ORGANISMS

The growth of filamentous organisms is influenced by several factors, which include operational conditions of an activated-sludge plant such as MCRT (or F:M), DO concentration, design of the plant, influent wastewater feeding pattern, influent wastewater characteristics, and temperature. These same factors also influence the growth of other organisms, including floc-forming organisms. Filamentous organisms become dominant when these factors provide conditions for faster growth rates of filamentous organisms than those of floc-forming organisms. Some filamentous organisms also become dominant when they are more resistant to starvation than floc-forming organisms.

MEAN CELL RESIDENCE TIME. Some filamentous organisms are observed at relatively high MCRTs but many other types are found at a wide range of MCRTs. Results of the studies conducted by Richard (1989) on activated-sludge treatment plants in Colorado are presented in Figure 8.17. Types 0041, 0675, 1851, 0092, and *M. parvicella* are all observed at high MCRTs such as more than 10 days. The other filamentous organisms found in activated-sludge plants operated at high MCRTs include types 0581, 0803,

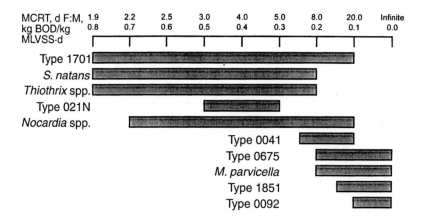

Figure 8.17 **Correlation between MCRT and occurrence of filamentous organisms in activated-sludge processes (modified from Richard, 1989).**

and 0961 (Wanner, 1994). Type 021N is observed at moderate MCRTs. *Nocardia* spp., *Thiothrix* spp., *S. natans*, and type 1701 are found at a wide range of MCRTs (Richard, 1989).

DISSOLVED OXYGEN CONCENTRATION. Some filamentous organisms are associated with low DO concentrations in the biological reactor because they grow faster than floc-forming organisms at low DO concentrations. The DO level at which filamentous organisms outgrow floc-forming organisms depends on F:M or MCRT. Type 1701, *S. natans*, *H. hydrossis,* and type 1863 occur in plants operated at low DO concentrations and low to moderate MCRTs. *Microthrix parvicella* is observed at low DO concentrations combined with high MCRTs

DESIGN OF BIOLOGICAL REACTORS AND INFLUENT WASTE-WATER FEEDING PATTERNS. Excessive growth of filamentous organisms depends on the design of the biological reactor, whether it is a completely mixed or a plug-flow reactor, or has initial unaerated zones such as anoxic or anaerobic zones where the RAS and the influent wastewater are mixed together. Filamentous organism growth is also influenced by influent wastewater feeding pattern, whether it is intermittent or continuous. Filamentous organisms whose growth are encouraged by continuously fed, completely mixed biological reactors include types 021N, 1851, 1701, *Thiothrix* spp., *S. natans*, *N. limicola*, *H. hydrossis*, and *Nocardia* spp. (Jenkins et al., 1993). The filamentous organisms that are observed in plants operated at high MCRTs with initial anoxic and anaerobic zones (such as biological nutrient removal plants) include *M. parvicella* and types 0092, 0041, and 0675 (Jenkins et al., 1993, and Wanner, 1994). Type 0914 also often occurs in high MCRT biological nutrient removal activated sludges (Jenkins et al., 1993). Design of biological reactors, secondary clarifiers, and scum removal systems strongly influences the level of foams caused by *Nocardia* spp., *M. parvicella*, and type 1863. These organisms float and remain on the surface of biological reactors and clarifiers. If there is no free outlet from the reactor to the clarifier and from the clarifier to the effluent, they are trapped and accumulate in the process. Recycling of centrates from dewatering and thickening centrifuges that contain *Nocardia* spp. may promote their growth in the biological reactor.

INFLUENT WASTEWATER CHARACTERISTICS. Characteristics of concern include the nature of organic substrate, whether it is soluble or particulate, whether it is readily or slowly biodegradable; nutrient concentration, such as nitrogen, phosphorus, or trace metals; sulfide concentration; and pH.

Nature of the Organic Substrate. Readily biodegradable substrates include the following compounds: glucose and glucose-containing di- and polysaccharides, alcohols (methanol is not included), volatile fatty acids, and amino acids, especially those containing sulfur in a molecule (Wanner, 1994). The

---

following filamentous organisms are capable of using these soluble, readily biodegradable substrates: *S. natans*, type 021N, *Thiothrix* spp., *H. hydrossis*, *N. limicola*, type 1851, and *Nocardia* spp. Filamentous organisms whose growth may be encouraged by slowly metabolizable substrates include types 0041, 0675, 0092, *M. parvicella*, and *Nocardia* spp. Some filamentous organisms grow well in the presence of organic acids. These include *Beggiatoa* spp., *Thiothrix* spp., and type 021N.

Nutrients. Nutrients that affect growth of filamentous organisms include nitrogen, phosphorus, trace elements, and growth factors. Deficiency in one or more of these nutrients may result in excessive growth of filamentous organisms. Type 021N, *Thiothrix* spp., type 0041, and type 0675 are all associated with nitrogen and phosphorus deficiency. Type 021N and *Thiothrix* spp. form rosettes and gonidia and contain PHB granules but no sulfur granules under nitrogen- or phosphorus-deficient conditions (Jenkins et al., 1993). Types 0041 and 0675 produce a Neisser-positive slime coating when nutrient deficiency occurs (Jenkins et al., 1993). Several authors (Ostrander, 1992, and Simpson et al., 1991) have reported the occurrence of bulking as a result of trace element and growth factor deficiency. Simpson et al. (1991) observed that an activated-sludge plant fed with influent wastewater containing sulfide concentrations of 1 to 15 mg/L experienced foaming and bulking as a result of excessive growth of *M. parvicella* and type 021N. When the plant was dosed with a metal ion solution, bulking and foaming were eliminated. Ostrander (1992) described a situation where foaming caused by *M. parvicella* was eliminated by adding a folic acid solution.

Sulfide Concentration. *Thiothrix* spp., type 021N, *Beggiatoa* spp., and type 0914 grow well at high concentrations of sulfide. They oxidize sulfide to elemental sulfur and deposit the sulfur as intracellular granules.

pH. Low pH (pH < 6.0) encourages the growth of fungi. When influent wastewater contains a strong acid, or nitrification occurs in plants treating low alkalinity wastewater, the pH of mixed liquor may decrease to less than 6.0. This condition coupled with high organic load may result in the excessive growth of fungi (Sezgin, 1989).

TEMPERATURE. Temperature influences the growth rate of filamentous organisms by affecting the DO levels in the mixed liquor by affecting the solubility of oxygen and the rate of metabolic processes. Richard et al. (1985) observed that bulking problems caused by low DO filaments, such as *S. natans* and type 1701, required higher DO levels to cure them as the temperature of the mixed liquor was increased. Jenkins et al., (1993) reported that, as the mixed liquor temperature increased, the washout MCRT for *Nocardia* spp. decreased. Wanner (1994) observed a shift in the dominance of *Nocardia* spp. during summer months to *M. parvicella* in winter months.

CONTROL OF FILAMENTOUS ACTIVATED-SLUDGE BULKING

Whenever filamentous activated-sludge bulking occurs, the following approach should be followed to rectify the problem. First, identify the filamentous organisms causing the bulking problems. Next, determine the plant operating conditions and changes in wastewater strength and composition. These include MCRT or F:M; biological reactor DO concentration; influent wastewater 5-day biochemical oxygen demand (BOD_5), nitrogen, phosphorus, and sulfide concentrations; septicity; and pH. Then, correlate the presence of particular filamentous organisms with plant operating conditions or wastewater composition. Finally, rectify the problem by making changes in plant operation, wastewater composition, chemical addition, or undertaking plant upgrading.

If, for example, bulking is caused by high levels of sulfides or septicity, influent wastewater chlorination, aeration, or any other action to reduce septicity may be initiated. If bulking is caused by nutrient deficiency, nutrient addition may be initiated. Control methodologies for filamentous activated-sludge bulking are as follows:

- Manipulation of plant operating conditions. These include manipulation of RAS flow rate, reduction in clarifier solids loading, and manipulation of DO concentration in the biological reactor.
- Chemical addition. The chemicals used are flocculating agents, such as polyelectrolytes, iron salts, aluminum salts, and lime, or toxic chemicals, such as chlorine, hydrogen peroxide, and ozone.
- Adjustments in wastewater composition by nutrient addition, sulfide removal, or pH adjustment.
- Modifications in the activated-sludge process.

MANIPULATION OF PLANT OPERATING CONDITIONS.
Manipulation of Return Activated Sludge Rate. This remedial action can be applied immediately. However, a decision to increase or decrease the RAS rate depends on the overload condition of the clarifier with respect to its clarification or thickening capacity. Detailed and time-consuming solids-flux curves are required for this type of analysis. For further information on secondary clarifier operating strategies, refer to Keinath et al. (1977), Keinath (1985), Daigger and Roper (1985), and Jenkins et al. (1993). However, simpler settling tests may serve as well. Comparing the settled sludge concentration to the actual return sludge concentration can serve as a guide. For further information, refer to WPCF (1987).

Reduction in Clarifier Solids Loading. This task can be accomplished in two ways: first, by reducing MLSS concentration through wasting more sludge and, second, by changing influent wastewater feed locations. However,

---

rectifying bulking problems by reducing the sludge mass may have some limitations. For example, wasting sludge to reduce the solids loading to the clarifier results in a reduction in MCRT. If nitrification is to be achieved, this method may result in the washout of nitrifiers from the process. Furthermore, if the bulking is caused by DO concentration, organisms such as *S. natans* or type 1701, increasing sludge wasting results in a higher F:M and lower DO concentration, which aggravates the problem further. However, if the bulking is caused by high-MCRT organisms such as types 0041, 0675, and 1851, decreasing the MCRT, in fact, results in elimination of these organisms. The process of MCRT reduction is a rather slow process and, in a completely mixed activated-sludge process, a period equal to three MCRTs is required to wash out filamentous organisms.

The second methodology to lower the solids loading to the clarifier involves changing the flow regime from plug flow to step feed by altering the influent wastewater feed locations.

Manipulation of Dissolved Oxygen Concentrations in the Biological Reactor. If microscopic examination indicates that bulking is caused by type 1701, *S. natans,* or *H. hydrossis* at low to moderate MCRTs or *M. parvicella* at high MCRTs, it is likely that low biological reactor DO concentration is the cause of the bulking. The bulking problem can be rectified by increasing the DO concentration in the biological reactor. The level to which the DO should be increased depends on the MCRT or F:M at which the plant is operated. For a given F:M, there is a critical DO level below which sludge bulking caused by low-DO filaments will occur. However, if the DO level is increased above the critical level, low-DO filaments are eliminated. Sezgin (1977) conducted laboratory experiments using completely mixed air and oxygen activated-sludge systems fed with settled domestic wastewater and found that, at an average F:M of 0.36 (based on chemical oxygen demand [COD] and volatile suspended solids [VSS]) and DO level of 1.3 to 1.8 mg/L, no bulking was observed. When the DO level was reduced to 0.3 to 0.9 mg/L at the same F:M, bulking occurred. Increasing the DO level from 0.3 to 0.9 mg/L back to 1.5 to 2.0 mg/L resulted in curing the bulking. Bulking caused by low-DO filamentous organisms can also be rectified by reducing F:M. Sezgin (1977) observed that when the activated-sludge process operated at a F:M was subjected to a higher F:M for a given biological reactor, DO concentration bulking caused by *Sphaerotilus* sp. occurred. When the F:M was reduced to previous levels, *Sphaerotilus* sp. was reduced to insignificant levels and bulking was cured.

Studies by Palm et al. (1980) also indicated the presence of a critical DO level for a given F:M at which bulking will occur. They observed that the higher the F:M, the greater the DO concentration necessary to prevent bulking caused by *Sphaerotilus* sp. The relationship between F:M and biological reactor DO concentration is presented in Figure 8.18 for a completely mixed, bench-scale activated-sludge system. However, this relationship is valid only for the particular wastewater and activated-sludge system described here and should not be applied universally to other plants. Palm et al. (1980) also observed that

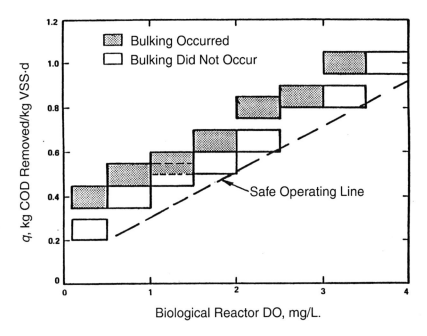

Figure 8.18 **Combinations of F:M and biological reactor DO concentrations that result in filamentous bulking and nonbulking activated sludges (Palm et al., 1980).**

bulking occurred faster relative to the curing period. Bulking because of low DO occurred in approximately 1.5 MCRTs. The period to cure the bulking by increasing biological reactor DO level took approximately 3 MCRTs.

Periodic aeration cures bulking caused by certain types of filamentous organisms. Bulking because of *S. natans* was cured in the South Charleston WWTP by on–off aeration (Schuyler, 1999). However, periodic aeration may not be effective in eliminating types 0041, 0675, 0092, and *M. parvicella* (Schuyler, 1999).

CHEMICAL ADDITION. The biological methods that cure bulking by washing out filamentous organisms require longer times than chemical methods. Biological methods may take up to 3 MCRTs to cure the bulking problem. On the other hand, addition of flocculating agents may result in immediate improvement in settleability of activated sludge whether the poor settleability of activated sludge is caused by excessive growth of filamentous organisms or the presence of high concentrations of polysaccharides (i.e., viscous bulking). However, with the addition of flocculating agents, neither the growth of filamentous organisms nor polysaccharide production are eliminated. Therefore, the problem may not be cured, just masked by chemicals. However, chemicals may allow temporary control of the process while other modifications are made to eliminate the filamentous growth or slime

---

Filamentous Organisms

production. Addition of toxic chemicals requires longer periods to cure the sludge bulking than the period required for the addition of flocculants.

Addition of Flocculating and Weighting Agents. Flocculating and weighting agents used to improve the settleability of activated sludge include polyelectrolytes, iron or aluminum salts, or lime [$Ca(OH)_2$]. Polyelectrolytes can be added to the effluent end of biological reactors, to mixed liquor channels after biological reactors, or to clarifier center wells. A cationic polymer or a combination of cationic and anionic, or a combination of metal ion (i.e., iron or aluminum) and anionic polymer can be used. Optimum dose should be determined using jar tests and should be verified periodically. Typical dosages of cationic polymers may vary between 1.6 and 2.8 g/kg TSS (1.6 and 2.8 lb/1000 lb) (Singer et al., 1968).

Iron salts used to improve activated-sludge settling include ferric chloride ($FeCl_3$), ferric sulfate [$Fe_2(SO_4) \cdot 3H_2O$], and ferric sulfate ($FeSO_4 \cdot 7H_2O$). A common aluminum salt for bulking control is alum [$Al_2(SO_4)_3 \cdot xH_2O$, $x = 13$ to 18].

Flocculating and weighting agents such as iron or aluminum may be applied to the influent of the activated-sludge process, to the biological reactor, or to the mixed liquor channel after the reactor. They can be added continuously or intermittently. Again, the optimum dose should be determined by performing jar tests. Typical dosage levels of iron salts are 7 mg iron/L based on biological reactor volume and 10 to 14 mg iron/L on the basis of influent flow and alum at 5 to 6 mg aluminum/L on the basis of influent flow for continuous application (Wanner, 1994). For intermittent addition for a period of at least 10 days, the dosages on the basis of biological reactor volume are 35 mg iron/L, 15 to 20 mg aluminum/L, and 10 to 15 mg calcium/L for iron salts, aluminum salts, and lime, respectively (Wanner, 1994). However, when these chemicals are added, additional chemical sludge is generated in the process.

Addition of Toxic Chemicals. Bulking can be cured by killing filamentous organisms selectively with the addition of toxic chemicals. However, it must be realized that, when sheathed filaments are involved, the organisms may die, but improved settling will not be observed until a significant amount of sheaths is wasted from the system. The chemicals used for controlling filamentous bulking include chlorine (Cl_2), hydrogen peroxide (H_2O_2), and ozone (O_3).

CHLORINE ADDITION. Chlorine is the most widely used and also the cheapest chemical to control bulking. If used properly, it does not have an adverse effect on the activated-sludge process. The critical dosage at which chlorine starts to interfere with the process depends on the type of process, whether it is nitrification, denitrification, or biological phosphorus removal. This critical dose also may vary with wastewater characteristics. Therefore, experimentation is needed to determine the critical dosage. Typical values of the critical dosage over which BOD_5 and TSS removal, soluble COD removal, nitrification, denitrification, and biological phosphorus removal are affected are 8 to 15, less than 8, less than 5, less than 5, and approximately 2 g Cl_2/kg MLSS·d, respectively (8 to 15, less than 8, less than 5, less than 5, and approximately 2 lb/d/1000 lb).

Filamentous organisms that are eliminated by chlorine addition include *S. natans*, *H. hydrossis*, types 021N, 1851, 1701, 0803, 0041, 0675, 0961, 0581, 0092, 1863, *Thiothrix* spp.; *N. limicola* II; *Nocardia* spp. (including *Rhodococcus* spp.); and *M. parvicella*. Some reports (Wanner, 1994) indicate that *M. parvicella* and type 0092 may be more resistant to chlorine compared to other filamentous organisms. Also, some filamentous organisms, such as *Nocardia* spp., may be more difficult to eliminate without killing the floc-forming organisms because they mostly grow within the floc.

Chlorine dosing point should be selected as a location where chlorine demand is minimum. The substances that exert a chlorine demand are ammonia, nitrite, sulfide, and dissolved and particulate organic matter. Chlorine should not be added at a point where the RAS mixes with the influent wastewater because the influent wastewater may contain high concentrations of ammonia, sulfide, and dissolved and particulate organic matter. These substances will react with chlorine, making it ineffective for killing filamentous organisms or higher chlorine doses are required to be effective for bulking control.

Chlorine can be dosed in several locations to rectify bulking problems, including the RAS line, the biological reactor, or a side stream that is pumped from the biological reactor effluent end and returned to the reactor influent (Jenkins et al., 1993). Among these dosing locations, the RAS line has been the most frequently used. The selection of the dosing location depends on several criteria. The dosing point should result in the targeted overall chlorine mass dose rate, chlorine concentration, and exposure frequency and prove to be a point of excellent mixing and minimum chlorine demand. The typical overall chlorine mass dose rate applied for bulking control ranges between 2 and 15 g Cl_2/kg SS·d (2 and 15 lb/d/1000 lb) (Jenkins et al., 1982). When chlorine is applied at the overall mass dose rate for bulking control, the chlorine concentration at the point of application should range between 5 and 10 mg/L (although a range of from 0.7 to 20 mg/L has been observed in the literature [Jenkins et al., 1982]). The frequency of exposure of biomass at the dosing point should be greater than three per day for effective bulking control (Jenkins et al., 1982).

Mixing should be adequate to dilute chlorine immediately. If mixing is not sufficient, some biomass will not be subjected to high chlorine concentration or may not be dosed at all. Typical chlorine application points with good mixing include locations following elbows in pipes, volutes of RAS pumps, RAS discharge points, and through a manifold into a waterfall.

The following procedure should be used when applying chlorine for bulking control:

(1) Choose a chlorine application point.
(2) Select a target SVI, for example, SVI = 150 mL/g.

(3) Plot a trend graph of the SVI daily and apply chlorine when the target SVI is significantly and consistently exceeded.

(4) If RAS chlorination is being used for the first time and the bulking problem does not pose a crises situation, start with a low chlorine dose such as 2 g Cl_2/kg MLSS·d (2 lb/d/1000 lb).

(5) Apply this dose for several days, check SVI and effluent turbidity, and conduct microscopic examination daily.

(6) If the chlorine dose does not reduce SVI, increase the chlorine dose gradually.

(7) Inspect secondary effluent for milky appearance, a significant increase in turbidity, and presence of white foam, which all indicate chlorine overdose.

HYDROGEN PEROXIDE ADDITION. Hydrogen peroxide acts in two ways to control filamentous activated-sludge bulking: it kills filamentous organisms in a similar fashion to chlorine and it provides an additional DO source, which may be effective in curing low-DO bulking. Hydrogen peroxide was applied in dosages of 0.1 to 0.4 g H_2O_2/kg MLSS·d (0.1 to 0.4 lb/d/1000 lb) for bulking control (Keller and Cole, 1973). Hydrogen peroxide was also used in concentrations of 20 to 200 mg/L on the basis of total volume of biological reactor and clarifier and 200 to 400 mg/L on the basis of influent wastewater flowrate to cure filamentous bulking. Hydrogen peroxide can be applied to the RAS line, biological reactors, and the mixed liquor channels between biological reactors and mixed liquor channel.

OZONE ADDITION. Ozone has recently been used for bulking control (van Leeuwen and Pretorius, 1988). When ozone is added to water, peroxide and hydroxide radicals are formed. It is believed that these radicals act as the disinfectants to control bulking. Ozone has been applied to the biological reactor in dosages of 4 to 6 g O_3/kg MLSS·d (4 to 6 lb/d/1000 lb) or 6 to 9 mg/L on the basis of influent wastewater flowrate (van Leeuwen and Pretorius, 1988).

ADJUSTMENTS IN WASTEWATER COMPOSITION. Nutrient Addition. Nutrient deficiency resulting from lack of nitrogen, phosphorus, trace elements, and growth factors has been reported in the literature (Jenkins et al., 1993; Ostrander, 1992; Simpson et al., 1991; Wanner, 1994; and Wood and Tchobanoglous, 1974). Bulking problems caused by a deficiency in nitrogen and phosphorus can be determined using a combination of mixed liquor microscopic examination and chemical analysis of influent wastewater, mixed liquor, and secondary effluent. If the microscopic examination provides the following information, nutrient deficiency from lack of nitrogen or phosphorus is suspected: (a) presence of excessive growth of *Thiothrix* spp. and type 021N; (b) extensive gonidia and rosette formation by *Thiothrix* spp. and type 021N; (c) positive India ink test; (d) presence of foam on the surface of biological reactors in the absence of *Nocardia* spp., *M. parvicella*, or type 1863; and (e) presence of Neisser-positive tetrad cells in the absence of an anaerobic–aerobic sequence during aeration and large amounts of intracellular PHB granules in the floc bacteria.

If the chemical analysis yields the following information, nutrient deficiency from lack of nitrogen and phosphorus is also suspected: (a) If BOD_5-to-nitrogen and BOD_5-to-phosphorus ratios in the influent to the biological reactor are greater than 20 and 100, respectively. These criteria, however, are not reliable. Some activated-sludge plants treating industrial wastewaters having these ratios greater than 20 and 100, respectively, have been operated without bulking problems; (b) If total Kjeldahl and total phosphorus contents of the organic fraction of mixed liquor solids are less than the range of 4.9 to 16.3% and 1.2 to 2.3%, respectively; (c) If carbohydrate content of mixed liquor solids is greater than 25 to 30% expressed as glucose; (d) If effluent soluble inorganic nitrogen and orthophosphate concentrations are less than 0.5 mg/L. Again, these criteria may not be reliable because some plants are operated at less than 0.5 mg/L orthophosphate in the effluent without bulking problems; and (e) If there is poor effluent quality (relatively low BOD_5 or COD removal).

Nitrogen and phosphorus deficiency not only cause filamentous activated-sludge bulking but also viscous bulking. However, the bulking problems can be rectified by adding supplemental nitrogen or phosphorus. The quantity of nutrients to be added varies with MCRT (or F:M), temperature, availability of nitrogen and phosphorus in the influent, and variations in the composition of influent wastewater (Sezgin, 1994).

Nutrient requirements decrease with increasing MCRT. Cell lysis occurs at high MCRTs, resulting in release of nutrients to the mixed liquor. Temperature affects nutrient requirements adversely. As the temperature increases, nutrient requirement per kilogram of BOD_5 removed decreases. This is because more of the BOD is used for cell maintenance energy, not for cell growth, as the temperature increases. On the other hand, nitrogen and phosphorus are required for cell growth, not for cell maintenance. Composition of nitrogen and phosphorus and carbon compounds influences the amount of nutrients to be added. If the nitrogen and phosphorus compounds are not in readily available form and the carbon source is an easily metabolizable substrate such as volatile organic compounds, supplemental nitrogen and phosphorus addition may be required to meet the nutrient requirement. Daily or seasonal variations in the carbon sources of the influent wastewater may also result in variations in nutrient requirements. In this case, nutrient addition should also be paced with the carbon source availability.

Various ammonia compounds such as anhydrous ammonia (NH_3), ammonium hydroxide (NH_4OH), ammonium bicarbonate (NH_4HCO_3), and ammonium chloride (NH_4Cl) can be used as nitrogen sources. The important point to remember when adding ammonia is that an overdose of ammonia may result in higher levels of nitrate-nitrogen (NO_3^-–N), in the biological reactor effluent from nitrification and sludge flotation in the secondary clarifiers because of denitrification. An overdose of ammonia may also result in high effluent levels of ammonia that eventually may cause fish toxicity in the receiving water. Phosphorus compounds that can be added as nutrient sources

---

Filamentous Organisms   *141*

include trisodium phosphate (Na_3PO_4), disodium phosphate (Na_2HPO_4), monosodium phosphate (NaH_2PO_4), monobasic ammonium phosphate ($NH_4H_2PO_4$), and phosphoric acid (H_3PO_4)(Sezgin, 1994).

After the addition of the required nutrient has started, effluent quality and settling characteristics of activated sludge should be monitored to determine the adequacy of nutrient addition. Effluent soluble inorganic nitrogen and orthophosphate levels should be in the range of 0.5 to 1.0 mg/L and may be as high as 3 mg/L for some types of readily biodegradable wastewater. An improvement in BOD_5 or COD removal and in sludge settling characteristics should be observed. If no improvements are observed, the dose of the deficit nutrient should gradually be increased until improvements are observed in effluent quality and sludge settling characteristics. If the nutrient dose selected originally improves both effluent quality and sludge settling characteristics, the dose can be lowered to help minimize chemical costs (Sezgin, 1994).

Sulfide Removal. The presence of high levels of *Thiothrix* spp., type 021N, *Beggiatoa* spp., and type 0914 may indicate the presence of high concentrations of sulfides in the influent wastewater. Total sulfide levels as high as 0.2 to 0.5 mg sulfur/L may result in sludge bulking by these organisms. High levels of *Thiothrix* spp. and type 021N may also indicate high levels of low-molecular-weight organic acids such as acetic acid. If septicity or organic acids originate from inefficient operation of primary clarifiers, optimization of clarifiers may help in reducing bulking. Otherwise, bulking caused by these organisms may be controlled by removing sulfides using one of the following methods. (a) Aeration. Influent wastewater can be aerated before entering the biological reactor to remove sulfides. A typical rate used for this purpose is approximately $12.7 \ m^3 \ air/m^3$ (1.7 cu ft air/gal) wastewater (Farquhar and Boyle, 1972). (b) Chemical addition. Chlorine, hydrogen peroxide, and metal salts are used to remove sulfide. In addition, chlorine and hydrogen peroxide act as disinfectants to kill filamentous organisms. Typical dosages of chlorine, hydrogen peroxide, and metal salts (e.g., iron salts) per milligram of hydrogen sulfide (H_2S) removed are 3 to 9 mg chlorine, 2.5 to 3 mg hydrogen peroxide, and 1 to 2 mg iron, respectively (Farquhar and Boyle, 1972, and U.S. EPA, 1985).

pH Adjustment. Low pH results in excessive growth of fungi. Bulking caused by fungi may be cured by increasing the pH of the biological reactor to neutral pH levels. Chemicals such as sodium hydroxide (NaOH), sodium bicarbonate ($NaHCO_3$), or lime can be added to the influent wastewater to raise pH. Sodium hydroxide addition does not result in additional chemical sludge, but sodium bicarbonate or lime addition creates additional chemical sludge.

MODIFICATIONS IN THE ACTIVATED-SLUDGE PROCESS.
Filamentous activated-sludge bulking may be rectified by modifying the flow pattern of the existing activated-sludge process. This can be accomplished by making changes in influent wastewater feeding points and feeding pattern or

installation of selectors ahead of the main biological reactor. The purpose of both measures is the same; that is, to create an environment that favors the growth of floc-forming organisms over filamentous organisms.

Changes in Influent Wastewater Feeding Points or Feeding Pattern. Activated-sludge processes that are completely mixed and continuously fed are more prone to bulking by low-F:M filamentous organisms than those with a plug-flow pattern or compartmentalized biological reactors or intermittently fed processes (Chudoba et al., 1973a and 1973b; Goronszy, 1979; Lee et al., 1982; Rensik, 1974; Rensik et al., 1982; and van den Eynde et al., 1982). In completely mixed and continuously fed biological reactors, influent waste-water is diluted immediately on entering the reactor, resulting in relatively low carbonaceous substrate concentrations. Some filamentous organisms can grow faster than floc-forming organisms under low carbonaceous substrate concentrations. Therefore, those filamentous organisms that grow faster can cause bulking in completely mixed activated-sludge systems. On the other hand, in plug-flow biological reactors, there is a carbonaceous substrate gradient along the reactor. The substrate concentration is high at the head of the reactor and low at the end of the reactor. In plug-flow biological reactors where the influent wastewater is mixed with RAS at the head of the reactor, the organisms are subjected to high substrate concentrations. Some floc-forming organisms not only grow faster than some filamentous organisms at high substrate concentrations, but can take up the excess substrate and store it and consume it later when the substrate concentration is depleted. For this reason, plug-flow reactors are less prone to sludge bulking by low-F:M filamentous organisms compared to completely mixed activated-sludge processes.

Intermittently fed biological reactors, compartmentalized reactors, batch reactors, and a process that incorporates a mixing tank (a selector), where RAS is mixed with influent wastewater ahead of the reactor, all produce a carbonaceous substrate gradient through the reactor or a high instantaneous substrate concentration at the mixing point followed by close-to-zero soluble substrate concentration in the reactor. Therefore, these systems are less prone to sludge bulking.

Sludge bulking resulting from low F:M organisms can be rectified in plug-flow biological reactors where RAS is returned to the head of the reactor and influent wastewater is fed as step feed along the reactor by changing the feed locations. For example, all of the influent can be redirected to the head of the reactor where it is mixed with RAS. Similarly, bulking in a sequencing batch reactor can be prevented by reducing the feeding period.

Installation of Selectors. A selector is a tank or a series of tanks where RAS is mixed with influent wastewater before the biological reactor (Figure 8.19). In the selector, an environment is created for floc-forming organisms that have the ability to take up the carbonaceous substrate and store it faster than filamentous organisms. These floc-forming organisms survive better than filamentous organisms by using the stored substrate in the subsequent

---

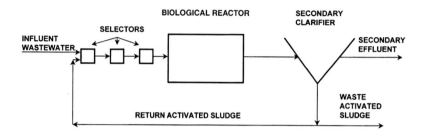

Figure 8.19   Activated-sludge process with selectors.

biological reactor environment where the soluble substrate concentration is close to zero. Furthermore, not only can the floc-forming organisms take up the available food faster, they can actually use the food within the environment provided, outgrowing the competing filaments. Depending on the type of the filamentous organism, the floc-forming organisms can grow faster in the selector than many filamentous organisms can grow. There are three types of selectors: aerobic or oxic, anoxic, and anaerobic.

AEROBIC OR OXIC SELECTORS. Typical DO concentration is in the range of 1 to 2 mg/L. In this type of selector, a portion of the carbonaceous substrate is oxidized in the presence of DO and the energy obtained from the oxidation process is used to transport and store the remaining carbonaceous substrate inside the cell. In the biological reactor, the stored substrate is oxidized to carbon dioxide and water in the presence of DO and the energy obtained during the oxidation process is used for making new cells from the stored substrate. Filamentous organisms that cannot store the soluble carbonaceous substrate faster than floc-forming organisms are eliminated from the process. Aerobic selectors should be designed to produce a selector soluble COD effluent of 60 mg/L to function properly and have at least three compartments (Jenkins et al., 1993). The organic loading in the first compartment should be 6 kg BOD_5/kg MLSS·d (Jenkins et al., 1993).

ANOXIC SELECTORS. In these selectors, DO is absent, but nitrate is present. In the selector, a portion of soluble carbonaceous substrate is oxidized and nitrate is reduced to nitrogen gas. The energy obtained in the oxidation process is used to transport and store the remaining carbonaceous substrate within the cell. In a biological reactor, the stored substrate is oxidized to carbon dioxide and water in the presence of DO and the energy obtained from the oxidation process is used for making new cells from the stored substrate. Filamentous organisms that cannot denitrify are eliminated from the process. Anoxic selectors should also be designed to produce a soluble effluent COD of 60 mg/L (Jenkins et al., 1993). An anoxic selector can be a single tank or three tanks in series. Typical denitrification rates in the anoxic selectors are 5 to 10 mg NO_3^-–N/g MLSS·h at 20 °C (68 °F) (Jenkins et al., 1993).

ANAEROBIC SELECTORS. In these selectors, DO and nitrate are absent, but inorganic polyphosphate granules within certain floc-forming microorganisms are present. In the anaerobic selector, polyphosphate granules are hydrolyzed and orthophosphate is released. The energy generated from the hydrolysis of polyphosphate granules is used to transport and store the soluble carbonaceous substrate within the cell. In the biological reactor, stored substrate is oxidized and the energy generated from the oxidation is used to take orthophosphate from the bulk medium and synthesize new polyphosphate granules and synthesize new cell material. Filamentous organisms that are not capable of carrying out these processes are eliminated from the system. The detention time in the anaerobic selector should be approximately 0.75 to 2 hours for the selector to be effective in controlling the filamentous organisms when treating domestic wastewaters and the soluble effluent COD from the selector should be 60 mg/L (Jenkins et al., 1993).

Some filamentous organisms can be eliminated from the process with the use of selectors, but others cannot. They are shown in Table 8.9. Selectors cannot cure filamentous bulking if bulking is caused by low pH, septic wastewater, high levels of sulfides, or nutrient deficiency (Jenkins et al., 1993).

CONTROL OF FOAM—SCUM FORMED BY NOCARDIOFORMS

TYPES OF FOAM–SCUM OBSERVED IN ACTIVATED SLUDGE.
Several different types of foam and scum are observed in activated-sludge plants. White, frothy foam may be observed during the startup of activated-sludge plants (Jenkins et al., 1993). After the process is established, the foam disappears. A different type of foam that also is white and frothy may appear

Table 8.9   Effectiveness of selectors against filamentous organisms (Jenkins et al., 1993).

Selectors effective	Selectors not always effective
Sphaerotilus natans	Type 0041
Type 1701	Type 0675
Haliscomenobacter hydrossis	Type 0092
Nostocoida limicola	*Microthrix parvicella*
Type 1851	
Type 021N[a]	
Thiothrix sp.[a]	
Nocardia sp.[b]	

[a] Selectors are not effective when caused by nutrient deficiency or septic wastewater.
[b] Aerobic and anaerobic selectors are not always effective. Anoxic selectors may be effective.

in activated-sludge plants when industrial wastewater containing slowly degradable surfactants is discharged. Sezgin (1989) observed that a white industrial foam appeared at the R.M. Clayton WWTP (Atlanta, Georgia) and covered the biological reactor, catwalks, and clarifiers. The foam did not cause any solids loss to the effluent and disappeared by itself. Foams may be produced in activated-sludge plants during episodes of viscous bulking caused by excessive production of extracellular polymers (Wanner, 1994). Foam caused by viscous bulking may contain significant amounts of biomass (Wanner, 1994). More information on this type of foam is provided under the section "Viscous Bulking." Other types of foam may occur in activated-sludge plants when an excessive amount of fine solids from solids handling processes are returned to the biological reactors within the recycled streams, such as anaerobic digester supernatant, centrifuge centrate, vacuum filter, or belt press filtrate (Jenkins et al., 1993). As a result, plants may experience turbid effluents (Jenkins et al., 1993). Use of high doses of chlorine may result in formation of white foam on clarifiers and on receiving waters and may produce highly turbid effluents. Scum may be observed on the surface of secondary clarifiers and anoxic tanks in nitrifying plants as a result of denitrification. Nitrates present in the effluent of biological reactors are converted to nitrogen gas, which attaches to sludge particles and makes them float in clarifiers and anoxic zones. Henze et al. (1993) evaluated the conditions leading to denitrification and concluded that 6 to 8 mg/L nitrate concentration in the effluent from biological reactors at 20 °C (68 °F) and a solids retention time of less than 1 hour in the bottom sludge layer of clarifiers might be sufficient enough to cause floating sludge.

In some treatment plants, a layer of heavy, viscous, brown scum develops over the biological reactors and secondary clarifiers. This layer, which may reach depths up to 2.4 m (8 ft), interferes with normal gas exchange at the air–water interface. Solids float over the weirs of the secondary clarifiers and may be lost from the system. The scum may cover catwalks and cause them to become slippery, requiring additional time and expense for frequent cleanup. Odor problems often develop. Scum also causes problems in anaerobic digesters, which includes blocking of gas mixing devices, covering top of digesters by escaping between the digester wall and floating covers, tipping of floating covers, fouling of the gas collection system, and reducing digester temperature by gas binding of sludge recirculation pumps.

Although the terms foam and scum are frequently used interchangeably, scum is probably more appropriate. In contrast to foam, scum cannot be dispersed easily by spraying with water.

Scum layers have been shown to be rich in actinomycetes, which include the genera *Nocardia*, *Gordona*, *Rhodococcus*, *Skermania*, and *Tsukamurella* (Goodfellow et al., 1998; Lechevalier, 1975; Sezgin et al., 1988; and Soddell et al., 1998). These organisms are collectively called nocardioforms because they exhibit typical nocardioform branching patterns (Soddell et al., 1998).

Microscopic examination of 250 bacterial strains isolated from 19 different treatment plants with scum problems indicated the presence of branched filaments as dominant organisms and they were identified as *Nocardia amarae*, *Nocardia rhodochrous* (now recognized as *Rhodococcus erythropolis* [Williams et al., 1984], *Nocardia asteroides*, *Streptomyces*, *Mycobacterium*, *Micromonospora*, and *Actinomadura* (Lechevalier, 1975). In the United States, the most frequently isolated organism has been *N. amarae* (Lechevalier and Lechevalier, 1974). Because of widespread occurrence of this organism, actinomycete scums have often been referred to as *Nocardia* scum or *Nocardia* foam. In reality, many other actinomycetes produce scum. Therefore, the term nocardioform will be used here to include other actino-mycetes. The term *Nocardia* will only be used when a reference is made to a study in which the term *Nocardia* was originally used. *Nocardia amarae* has recently been assigned to the genus *Gordona* as *Gordona amarae* (Klatte et al., 1994). Another nocardioform organism that is common in Australian WWTPs is *Nocardia pinensis* (Blackall et al., 1989). This organism has also been reclassified as *Skermania piniformis* (Chun et al., 1997). In Europe, the most frequently observed organism is *Rhodococcus* (Lemmer, 1989).

Scum caused by other morphologically different organisms has also been observed (Sezgin and Karr, 1986). A scum that is light brown and less viscous than that caused by *G. amarae* is generated by a rod-shaped organism. This organism typically occurs before the appearance of *G. amarae* and may also be observed with *G. amarae*. A survey of plants in Australia has indicated that scum has also been caused by coryneform-type organisms exhibiting little or no branching (Seviour et al., 1994). Another type is caused by an actino-mycete with longer branches than *G. amarae*. This scum is more viscous and has an orange-brown color. Scum also is produced by the filamentous organisms, type 1863 and *Microthrix parvicella* (Blackbeard et al., 1986). Some strains of type 1863 are now recognized as species of the genus *Acinetobacter* (Wagner et al., 1994). *Microthrix parvicella* has recently been determined to be an actinomycete and is distantly related to members of the mycolic acid containing actinomycetes (Blackall et al., 1994 and 1996).

ROLE OF NOCARDIOFORMS IN SCUM FORMATION. Scum forma-tion in activated-sludge processes occurs as a result of interactions of air, wastewater, and nocardioform organisms in the mixed liquor. Nocardioform organisms have hydrophobic cell surfaces. In other words, the cells of nocardioform organisms lack affinity for water and they concentrate at air–water interfaces, such as at the edges of air bubbles in wastewater. When mixed liquor containing nocardioform cells is aerated, the cells attach to air bubbles and the bubbles move upward to the surface of the biological reactor. In addition, nocardioform cells carry activated-sludge flocs with them to the biological reactor surface. At the surface, water drains from the scum and the scum becomes more concentrated in nocardioform organisms and activated-sludge flocs. Total suspended solids concentration of scums on biological reactors may range between 4 and 6%. Studies by Vega-Rodriquez (1983) indicated that, when activated sludge containing nocardioform organisms is aerated in the laboratory, 75 to 90% of the nocardioform filaments were transferred to the scum.

Hydrophobic surface characteristics of nocardioform organisms are thought to be caused by the presence of large amounts of mycolic acids (Bendinger et al., 1993; Lechevalier, 1986; and Mori et al., 1988), which are fatty acids. However, other studies (Stratton et al., 1998, and Sunairi et al., 1997) indicated that *R. rhodochrous* cells (which are nocardioform organism cells) contained no detectable mycolic acids when grown on Tween 80 at 10 and 17 °C (50 and 63 °F), but were extremely hydrophobic and produced a stable scum. Therefore, it was concluded that mycolic acid composition of cells of nocardioform organisms appeared to have no significant effect on cell surface hydrophobicity (CHS) (Stratton et al., 1998). On the other hand, CHS was found to change with culture age, growth temperature, and carbon source (Stratton et al., 1998).

The presence of surfactants increases both the amount of scum produced by an activated sludge containing nocardioform organisms and the stability of scum (Jenkins et al., 1993). The surfactant should be slowly biodegradable to exert this effect. Otherwise, it will be biodegraded quickly in the biological reactor. In the absence of a surfactant, a bubble formed by air on the reactor surface is rather unstable. When the water layer surrounding the bubble is drained, the layer becomes thinner and the bubble eventually bursts. The addition of a surfactant to a wastewater makes the water layer surrounding the bubble tougher. Therefore, it takes longer to burst the bubble and the foam formed by the bubbles becomes more stable.

The surfactants may be present in wastewater or may be produced by activated-sludge organisms. Khan and Forster (1988, 1990a, and 1990b) reported that *Rhodococcus rubra*, which is a nocardioform organism, produced surfactants and the surfactants extracted from the cells of *R. rubra* were capable of inducing foam in nonscumming activated sludges. Kilroy and Gray (1992) observed that nocardioforms proliferated and formed scum on the surface of biological reactors after the spillages of the surfactant ethylene glycol into the wastewater.

RELATION BETWEEN SCUM FORMATION AND PLANT OPERATION AND DESIGN. A literature survey suggests that the occurrence and proliferation of nocardioform organisms has been correlated with a number of plant operation and design conditions. Nocardioform organisms are typically observed at high MCRTs. Therefore, they can be eliminated by lowering MCRT through wasting more sludge (Milwaukee Mystery, 1969, and Pipes, 1978). Nocardioform organisms were observed in the biological reactors of the Jones Island Wastewater Treatment Plant, Milwaukee, Wisconsin (Milwaukee Mystery, 1969). The scum was eliminated by reducing the MLSS concentration, and hence the MCRT. Pipes (1978) reported that the scum caused by nocardioform organisms was observed at the Chicago Heights plant of the Bloom Township Sanitary District in Illinois and eliminated also by reducing the MLSS, and hence the MCRT. Sezgin and Karr (1986) described the appearance of nocardioform scum at the R.M. Clayton Plant in Atlanta, Georgia, within 3 weeks when the MCRT was increased from approximately 4 days to 12 days. When the MCRT was reduced to fewer than 3 days, the scum diminished within 4 days and was gone by 23 days. Wilson et al. (1984) reported elimination of nocardioform scum by reducing the MCRT to approximately 1 day at the 23rd Avenue treatment plant in Phoenix, Arizona.

Pipes (1978) surveyed 32 WWTPs and reported that nocardioform organisms were never observed in plants operating at less than 9-day MCRT. The plants that experienced scum problems also all had wastewater temperatures greater than 18 °C (64 °F). The effect of MCRT on the growth of nocardioform organisms at various temperatures was examined by Pitt and Jenkins (1990) and Cha et al. (1992) using laboratory-scale, activated-sludge systems. Pitt and Jenkins (1990) found that *Nocardia* counts of the mixed liquor fed with the wastewater from the Southeast Water Pollution Control Plant in San Francisco, California, increased with increasing MCRT from approximately 1.5 days to 20 days at temperatures ranging from 18 to 25 °C (64 to 77 °F). When the temperature was 13 °C (55 °F), *Nocardia* counts of the mixed liquor did not increase but were constant with the increasing MCRT from 1.5 days to 20 days. Cha et al. (1992) also reported that *Nocardia* counts of the mixed liquor fed with the wastewater from the Sacramento, California, Regional Wastewater Treatment Plant increased with increasing MCRT from approximately 1.5 days to 15 days at temperatures of 16 and 24 °C (61 and 75 °F).

Mixed liquor pH has a significant effect on the growth of nocardioform organisms. Hiraoka and Tsumura (1984) reported that the growth rate for *Nocardia amarae* (*G. amarae*) at pH 6.5 was approximately 66% of that at pH 7.0 and decreased to less than 1% at pH 5.0. Cha et al. (1992) observed that the optimum pH for growth of nocardioforms in a completely mixed laboratory-scale, activated-sludge process operated at MCRTs of 3 and 8 days was 6.5. The pH range studied in this investigation was between 6.5 and 7.5. *Nocardia* counts in the mixed liquor also were greater than in the system operated at an MCRT of 8 days compared to those in the system operated at 3 days for mixed liquor pH ranging from 6.0 to 7.0.

The growth of nocardioforms is enhanced when biological reactors and clarifiers are designed with submerged outlets. Cha et al. (1992) compared *Nocardia* counts in two laboratory-scale, activated-sludge systems. One of the systems had a scum-trapping configuration that included a subsurface mixed liquor flow from the biological reactor to the clarifier and a scum baffle in front of the secondary clarifier weir. The other activated-sludge system had a mixed liquor overflow from the biological reactor to the clarifier and no secondary clarifier scum baffle. Both systems were operated at MCRTs of 5 and 10 days. *Nocardia* count in the mixed liquor of the system with a scum-trapping configuration was five times greater than the *Nocardia* count in the system with no scum-trapping configuration at both MCRTs. In this study, scum and activated sludge were both wasted in the same proportions. However, in full-scale, activated-sludge plants with submerged outlets, sludge is wasted from the RAS line without wasting any from the scum. This practice would result in more scum accumulation on the surface of biological reactors and clarifiers. Once nocardioforms are on the surface, they have selective access to substrates that float. Such substrates include oils, fats, and greases. Nocardioforms are capable of using these substrates and other slowly biodegradable, high-molecular-weight compounds, including polysaccharides, proteins, pesticides, and aromatic compounds, and readily biodegradable compounds such as sugars and low-molecular-weight fatty acids (Lemmer and Kroppenstedt, 1984). Nutrients tend to accumulate at the air–water interface at concentrations as much as 100 times greater than in the water itself (Marshall, 1979). By producing surface-active agents that cause scum formation, nocardioforms are able to converge in a microzone of high nutrient concentration. Scum formation, therefore, represents a unique mechanism that allows the nocardioforms to outgrow their competitors. In the absence of nutrients, nocardioforms also have an advantage. One trait is their ability to survive without nutrients for long periods (several weeks), compared with other bacteria (several hours to a few days) (Boylen and Mulks, 1978). Consequently, long periods of starvation are likely to favor nocardioforms. All of this information suggests that a scum environment on the surface of biological reactors favors the growth of nocardioforms.

The growth of nocardioforms in biological reactors is also affected by in-plant practices used to deal with the scum removed from the reactors. If the removed scum is returned to the headworks, scum formation in biological reactors is enhanced because of the seeding of the influent with nocardioforms. If the scum is discharged to solids-disposal facilities, depending on the extent of treatment, seeding of the biological reactor through recycle flows with nocardioforms may not be as dramatic compared with the practice in which scum is returned to the headworks.

CONTROL AND PREVENTION OF SCUM FORMATION. The nocardioforms possess a number of traits that enhance their survival potential in biological wastewater treatment processes. Prevention of scum formation ultimately depends on controlling the specific set of conditions that favor these traits. For a given situation, experimentation with various control measures is needed to determine what is most effective. In some cases, a combination of methods may be required. A survey conducted by Pitt and Jenkins (1990) indicated that a number of methods have been used to control or prevent nocardioform scum in biological reactors. These control measures have been used with varying degrees of success (Table 8.10). In the following sections, currently available scum control methodologies are discussed.

Manipulation of Mean Cell Residence Time. Nocardioforms can be eliminated from activated-sludge processes by reducing MCRT (Cha et al., 1992; Milwaukee Mystery, 1969; Pipes, 1978; Pitt and Jenkins, 1990; Sezgin and Karr, 1986; and Wilson et al., 1984). The MCRT at which they can be washed out depends on the type of nocardioform organism, activated-sludge process design, and environmental conditions such as temperature and pH. Therefore, the operator has to experiment to determine the washout MCRT for nocardioform organisms in the plant. The pilot-plant and full-scale studies conducted by Cha et al. (1992) indicated that the washout MCRT for nocardioforms varied with mixed liquor temperature. Nocardioforms were washed

Table 8.10   Methodologies used to control biological scum in the United States (Pitt and Jenkins, 1990).

Scum control methodology	Number of plants using methodology	Success rate, %
Mean cell residence time	44	73
Mixed liquor or return activated sludge chlorination	48	58
Use of water sprays	58	88
Use of antifoaming agents	35	20
Reduced aeration	5	60

out from the biological reactors at an MCRT of 2.2 days at 16 °C (61 °F). When the temperature increased to 24 °C (75 °F), the washout MCRT was reduced to 1.6 days. Orange County (California) Sanitation District reported similar results in controlling *Nocardia* growth using MCRT (Ooten, 1989). The MCRT in the air activated-sludge plant is maintained at fewer than 2.7 days at an MLSS concentration of 900 mg/L with a temperature range of 21 to 31 °C (70 to 88 °F). The MCRT in the oxygen activated-sludge plant is maintained at fewer than 1.65 days at an MLSS concentration of 1000 to 1200 mg/L with the wastewater temperature between 21 and 31 °C (70 and 88 °F). Washout MCRT in the oxygen activated-sludge plant is lower than that of the air activated-sludge plant. The reason for this may be the occurrence of scum trapping and lower mixed liquor pH in oxygen activated-sludge plants compared to those in air activated-sludge plants (Jenkins et al., 1993). Optimum growth for nocardioforms is reported to occur at a pH of approximately 6.5. The pH of the mixed liquor in oxygen activated-sludge plants is also approximately 6.5. But, the mixed liquor pH in air activated-sludge plants is typically greater than 6.5 (Cha et al., 1992, and Jenkins et al., 1993). Jenkins et al. (1993) predicted that oxygen activated-sludge plants with a pH of 6.5 would have 20% more *Nocardia* than air activated-sludge plants with a pH of 7.0. The biological reactors in oxygen activated-sludge plants always have submerged mixed liquor outlets so that scum trapping on their surface occurs. Scum trapping in an activated-sludge system resulted in five times more *Nocardia* count in the mixed liquor compared with that in an activated-sludge system without scum trapping (Cha et al., 1992).

When scum trapping occurs, it is more difficult to wash out nocardioforms from the activated-sludge process because sludge is typically wasted from the RAS line, not from the scum on the surface of the biological reactors.

Although MCRT reduction is an effective method in eliminating nocardioforms in activated-sludge plants, it may not be a viable method for plants required to achieve nitrification or biological phosphorus removal. Nitrifying bacteria grow slowly. They, therefore, are observed at relatively high MCRTs. In fact, their MCRTs are greater than those MCRTs at which nocardioform actinomycetes are observed (Jenkins et al., 1983). When MCRT is reduced to wash out nocardioform organisms, nitrifiers are also washed out, resulting in high effluent ammonia levels. Reports indicate that washout MCRT for nocardioforms are also lower than those required for biological phosphorus removal (Jenkins et al., 1993; Mamais and Jenkins, 1992; and Shao et al., 1992).

Use of Selectors. The growth of nocardioform actinomycetes can be controlled by certain types of selectors. Cha et al. (1992) conducted laboratory-scale, activated-sludge experiments to determine whether *Nocardia* spp. can be eliminated by oxic selectors. They found that the effectiveness of the oxic selectors depended on the MCRT at which the activated-sludge systems were operated. When the MCRT was 5 days, the oxic selector effectively controlled the growth of *Nocardia* spp. However, when the MCRT was 10 days, the oxic selector was not effective in controlling the growth of *Nocardia* spp. Jenkins et al. (1993) reported that similar results were obtained in controlling *Nocardia* growth with oxic selectors in a full-scale treatment plant. The Upper Occoquan Sewage Authority Water Reclamation Plant in Centreville, Virginia, used oxic selectors to control filamentous bulking. The selectors were also effective in controlling *Nocardia* scum in summer as long as the MCRT was kept between 5 and 8 days, but not effective when the MCRT was greater than 10 days. In winter, *Nocardia* scum was not frequently observed even though the MCRT ranged between 10 and 15 days. Kappeler et al. (1993), however, was not able to control scum caused by both nocardioform organisms and *M. parvicella* using oxic selectors.

The effectiveness of anoxic selectors in controlling the nocardioform scum was studied by Sezgin and Karr (1986) at Utoy Creek Wastewater Treatment Plant, Atlanta, Georgia. The plant had two identical nitrifying biological reactors, each of which consisted of four passes. The first pass of one of the reactors was converted to an anoxic zone by reducing the airflow to the coarse bubble diffusers. The other reactor was used as a control. The results indicated that the biological reactor with an anoxic zone was capable of reducing scum coverage on the reactor and clarifiers on three occasions out of five compared with the control system. Because the aeration system was used to prevent the mixed liquor solids from settling in the first pass, the DO concentration was in the range of 0.2 to 0.5 mg/L in this pass. Gasser (1987) also reported that the *Nocardia* scum was successfully controlled at the Redding Regional Wastewater Treatment Facility in Redding, California, by shutting off aerators in biological reactors for a designated period of time. This practice resulted in anoxic periods in the reactors. He also reported that maintaining a DO concentration less than or equal to 0.8 mg/L from 12:00 p.m. to 9:00 p.m. each day moderately controlled the *Nocardia* scum (Gasser, 1987).

The effectiveness of anoxic selectors in controlling *Nocardia* scum was demonstrated by Cha et al. (1992). Two laboratory-scale, activated-sludge systems with and without an anoxic selector were operated side by side at an MCRT of 12 days. The *Nocardia* counts in the system with an anoxic selector were typically lower than those in the system without the anoxic selector. Albertson and Hendricks (1992) also achieved control of *Nocardia* more than

95% of time using anoxic selectors at the Phoenix 23rd Avenue Wastewater Treatment Plant in Phoenix, Arizona. The plant was operated at MCRTs between 5 and 12 days and consisted of three-compartment anoxic zones ahead of a plug-flow-type biological reactor. Kappeler et al. (1993) reported that anoxic selectors were effective in controlling *Nocardia* scum in full-scale, activated-sludge plants in Switzerland, but partially effective for the control of *M. parvicella* scum. The effectiveness of selectors in controlling the growth of nocardioform actinomycetes is reduced if the activated-sludge process has natural traps and baffles in the biological reactors so that scum is trapped or nocardioform actinomycetes are recycled back to the reactors (Albertson, 1992, and Jenkins et al., 1993).

Pure culture studies by Blackall et al. (1991) demonstrated that *N. amarae* neither grew nor was able to take up substrate (acetate was used as substrate) under anaerobic conditions. These results would indicate that scum caused by *N. amarae* could be controlled by anaerobic selectors. Therefore, anaerobic selectors were used to control *Nocardia* growth in both laboratory-scale and full-scale experiments by Pitt and Jenkins (1990). In most cases, only small and inconsistent reductions in *Nocardia* count were observed. On the other hand, the use of anaerobic selectors in controlling actinomycete scum was successful in Japan. Mori et al. (1992) reported that the foam coverage on the surface of the anaerobic–oxic channel ranged between 11 and 15% compared with 54 to 58% coverage on the conventional, fully aerated channel during the 18-month study period at Okinawa Wastewater Treatment Plant in Japan.

Use of Chemicals. The chemicals used in controlling the growth of nocardioforms include chlorine and ozone. Other chemicals tested for scum control are ferric chloride, aluminum salts, surfactants, polymers, and antifoaming agents.

CHLORINE ADDITION. Chlorine has been added in several locations for the control of nocardioform actinomycete scum. Chlorine addition points include the RAS line, mixed liquor, and onto the scum on the surface of biological reactors. The effectiveness of chlorinating RAS lines to destroy nocardioform actinomycetes has varied (Albertson, 1992; Jenkins et al., 1993; Sezgin and Karr, 1986; Wong and Chung, 1993; and Zickefoose and Vass, 1984). Sezgin and Karr (1986) reported that several attempts had been made in the city of Atlanta, Georgia, treatment plants to eliminate nocardioform actinomycete scum by chlorinating the RAS line. However, these attempts were unsuccessful because the dosages that were effective for scum control resulted in floc dispersion and effluent deterioration. Albertson (1992) also reported that addition of 1800 kg/d chlorine into the RAS line at the Phoenix (Arizona) 23rd Avenue Wastewater Treatment Plant was not successful in controlling nocardioform actinomycete growth. On the other hand, wastewater personnel in Stamford, Connecticut, had success in controlling scum by adding 10 mg/L chlorine to RAS lines for 36 to 48 hours (Sezgin and Karr, 1986). Wong and Chung (1993) reported that adding hypochlorite solution to the RAS line at a concentration of 10 mg/L reduced *Nocardia* counts within a few days in a bench-scale, activated-sludge system. Zickefoose and Vass (1984) also

---

reported some success in scum control with chlorine dosages of 5 to 7 kg chlorine/1000 kg (5 to 7 lb/1000 lb) solids added to either mixed liquor or return sludge. *Nocardia* scum in the first stage of the two-stage, activated-sludge process at the San Jose/Santa Clara, California, wastewater treatment plant was controlled by a combination of sludge wasting and RAS chlorination (Beebe, 1983). In this plant, F:M was increased from 0.35 to 0.5 kg BOD_5/kg MLVSS·d (0.35 to 0.5 lb/d/lb) by sludge wasting and the RAS line was chlorinated at approximately 4 kg Cl_2/10 kg MLVSS·d (4 lb/d/10 lb) for a period of 4 to 5 days (Beebe, 1983).

A better way of controlling scum is to spray chlorine solution or sprinkle powdered high-test calcium hypochlorite directly on the scum on the surface of biological reactors (Semon, 1985). This method of scum control has been applied in several plants in the United States. Albertson (1992) described the application at the Phoenix (Arizona) 23rd Avenue Wastewater Treatment Plant. Spray hoods were placed across the end of the third pass of the four-pass biological reactor. Chlorine solution containing 2000 to 3000 mg Cl_2/L was sprayed at a rate of 10 L/min on the scum. The hoods were used to prevent chlorine fumes from spreading in the surrounding environment. The scum was eliminated in 2 days. The chlorine concentration based on the influent wastewater flowrate was 0.5 to 1.0 mg Cl_2/L. During three subsequent incidents, scum was eliminated in between 1 and 1.5 days.

An important aspect of controlling scum formation is preventing reseeding of the mixed liquor through recycling of nocardioform-actinomycete-containing solids and supernatant. This can be accomplished by chlorinating recycle streams.

OZONE ADDITION. The effect of ozone on the growth of nocardioform actinomycetes was studied in pilot- and full-scale treatment plant trials (Goi et al., 1995, and Okouchi et al., 1996). Ozone was added directly to the first compartment of the three-compartment biological reactor of a pilot-plant oxygen activated-sludge process (Goi et al., 1995). Ozone concentration was varied between 2 and 6 mg/L on the basis of influent flowrate. The results indicated that ozone in the concentration range of 2 to 6 mg/L suppressed the growth of *Nocardia* spp. relative to those observed in the control system, which did not receive any ozone. Full-scale trials by Okouchi et al. (1996) revealed that ozone levels greater than 1.2 mg/L (based on influent flowrate) in the biological reactor were sufficient to control *Nocardia* growth. Furthermore, ozone levels as great as 3 mg/L did not have any adverse effects on effluent quality.

IRON AND ALUMINUM SALTS ADDITION. Duchene and Pujol (1991) observed that addition of ferric chloride at a dose of 4 g $FeCl_3$/kg MLVSS·d reduced nocardioform actinomycete scum within 2 weeks. The addition of aluminum salts also had similar results. However, these chemicals do not eliminate nocardioform actinomycetes, but flocculate them. Therefore, when their addition is stopped, nocardioform actinomycetes reappear.

SURFACTANT ADDITION. Pagilla et al. (1996) tested the effectiveness of a biodegradable non-ionic surfactant in reducing *Nocardia* scum in a bench-scale, activated-sludge process. Addition of 0.5 mg/L surfactant reduced *Nocardia* levels from 7.7×10^6 intersections/g VSS to 0.01×10^6 intersections/g VSS.

POLYMER ADDITION. Shao et al. (1997) reported successful control of *Nocardia* scum at the Terminal Island Treatment Plant in Los Angeles, California, with the addition of a polyacrylamide cationic polymer. The polymer was added at a concentration of 0.5 mg/L to the mixed liquor channel before the final clarifiers. After 3 days of polymer addition, the scum on the biological reactors and clarifiers was significantly reduced. Nocardioform actinomycetes, however, were not eliminated. Polymers incorporate nocardioform filaments to the activated-sludge flocs so that they are not carried to the surface of the biological reactor to form a scum layer. Once nocardioform filaments are incorporated to the flocs, they can be wasted from the system with the rest of the sludge.

ANTIFOAMING AGENT ADDITION. In bench-scale studies, a polyglycol antifoaming agent was shown to be effective for foam reduction (Lechevalier, 1975). A dosage of 0.1 mL/L mixed liquor was required. The actinomycete, however, was not eliminated.

Bioaugmentation. In theory, organisms that outcompete the actinomycetes for nutrients, and thereby interfere with their ability to form scum, could be used for scum control and prevention. However, such organisms must be able to survive in the treatment plant environment. Scum may be controlled by adding specialized bacteria (bioaugmentation) that release enzymes to degrade the lipids produced by the scum-forming bacteria. The nocardioform actinomycetes, however, are not destroyed. Several attempts to reduce scum formation by adding mutant bacteria for periods as longs as 1 month, and attempts to obtain a microorganism that would degrade *N. amarae* or a bacteriophage that would infect it, were all unsuccessful (Lechevalier, 1975, and Sezgin and Karr, 1986).

The effect of bacteria and enzyme (B–E) additives on *Nocardia* scum was also investigated by Franz and Matsche (1994). The B–E additive was tested in two plants. In one of the plants, the additive resulted in the replacement of *Nocardia* by *M. parvicella*, with no apparent improvements in the scum problems. The second plant had two parallel identical biological reactors. The B–E additive was added to one of the reactors at a dose of 15 g B–E additive/kg fat (or 15 kg B–E additive/d). The other reactor served as a control and received no B–E additive. The results indicated that the effluent quality from the two parallel reactors with respect to total organic carbon (TOC), COD, and nitrogen parameters did not differ. The fatty acid composition and grease content of the mixed liquor from both reactors also did not differ and the reactor treated with the B–E additive had more scum compared with the reactor with no B–E additive.

Termination of Preferred Food Sources. Many nocardioform actinomycetes use a variety of recalcitrant (difficult to degrade) hydrocarbons, including petroleum, fats, oils, and grease, as sources of carbon for growth. Because the presence of these compounds provides a selective advantage for nocardioform actinomycetes, limiting the introduction of these compounds to the treatment plant influent from industrial sources has been recommended to prevent scum formation (Gerardi, 1986). At the Flint River plant (Georgia), decreasing the concentrations of hydrocarbons in the influent from 200 to 500 mg/L down to 100 mg/L also reduced the scum problem (Sezgin and Karr, 1986). However, it cannot be concluded that the hydrocarbons were the sole cause of scum problems because scum formation has been reported at facilities treating industrial wastewaters lacking these compounds.

Scum Disposal. Nocardioform actinomycetes float on the surface. Therefore, they are not removed from the biological reactor as rapidly as the microorganisms suspended in the mixed liquor because the sludge wasting is typically carried out from the RAS line. However, they can be removed from the reactor surface selectively. Once they are removed, they should not be recirculated to the reactor. Pretorius (1987) recommended the use of a flotation tank in which removal of nocardioform actinomycetes was encouraged by means of vigorous aeration. Once they were floated, they could be removed from the surface of the flotation tank. Pretorius and Laubscher (1987) investigated the effectiveness of such a flotation tank in controlling scum in biological reactors. Two activated-sludge systems, each with a biological reactor volume of 200 L were run side by side. One of the systems incorporated a 16.7-L fine bubble flotation tank between the biological reactor and the secondary clarifier and the other system did not have the flotation tank, so served as a control system. After approximately 2 days of operation, the scum level in the system with the flotation tank was reduced to zero and the control system continued to have high scum levels. Full-scale trials of selective scum wasting were conducted by Jones (1988) at the Utoy Creek Wastewater Treatment Plant, Atlanta, Georgia. This plant had two identical four-pass biological reactors that could be operated in the plug-flow or step-feed mode. Each biological reactor had an aerobic digester attached to its first pass. The aerobic digester consisted of two sections: a reactor in which sludge was digested and a concentrator where digested solids were settled. Under normal operating conditions, settled digested solids were pumped to the dewatering centrifuges and the supernatant overflowed into the first pass of the biological reactor from the concentrator. The supernatant level in the concentrator could actually be regulated by the solids pumping rate. When the plant encountered scum accumulation on the biological reactors, the airflow to the first pass was first increased to strip nocardioform actinomycetes from the mixed liquor to the surface and then the supernatant level in the concentrator was lowered so that scum could overflow into the concentrator. Because scum was 2 to 3% in solids, it was directly pumped to the dewatering centrifuges from the concentrator and then incinerated. The findings of Jones (1988) are summarized by Richards et al. (1990). Scum hoppers were also installed at the effluent end of the biological reactors at R.M. Clayton Wastewater Treatment Plant at Atlanta, Georgia, to remove scum from the surface. Once removed, the scum is pumped to the anaerobic digesters. However, the plant has not been successful with scum removal because of problems associated with scum pumping, ineffective scum removal by the hoppers, and excessive digester foaming caused by the scum. As a result, walkways between the biological reactors are frequently covered with the scum.

Reduced Aeration Rates. The scum level is directly related to the vigor of aeration (Sezgin and Karr, 1986). *Nocardia amarae* produces a surface-active agent that induces foaming (Cairns et al., 1982, and Lechevalier, 1975). As the aeration is increased, more actinomycetes are stripped from the mixed liquor and trapped in the air bubbles of the scum. Therefore, lower aeration levels reduce scum formation. Caution must be exercised, however, to prevent a reduction of treatment efficiency.

Return of Anaerobic Digester Supernatant. A comparison of the operating records of several different plants showed that those plants that disposed of the supernatant from the anaerobic digester by returning it to the primary treatment system, without chlorination, did not have scum problems (Lechevalier, 1975). Plants that had a scum problem either lacked an anaerobic digester or did not return the digester supernatant to the primary treatment system. In addition, anaerobic digester supernatant was found to be toxic to *N. amarae* growing in Czapek's medium at $1:10^5$ to $1:10^6$ dilution. The toxicity was associated with the solids in the digester supernatant and could not be extracted with organic solvents. Returning the supernatant from the anaerobic digester to the treatment system was suggested as a method for controlling scum formation caused by *N. amarae*.

At four different treatment plants, scum control was attempted by returning anaerobic digester supernatant to the primary treatment system (Lechevalier, 1975). No anaerobic digester supernatant was added until scum formation began. This approach was only partially successful, possibly because of the loss of the solids from the supernatant in the primary treatment system. Returning the digester supernatant directly to the biological reactor was suggested as a more effective approach. During a test of 14 strains of actinomycetes, only *N. amarae* was found to be sensitive to anaerobic digester supernatant (Lemmer and Kroppenstedt, 1984). On the other hand, studies by Wheeler and Rule (1980) indicated that the addition of 1 to 5% (v/v) digester supernatant did not reduce *Nocardia* foam and predatory organisms were not effective.

Addition of Mixed Liquor Supernatant from an Offline Biological Reactor. One suggested method for controlling scum formation is the addition of mixed liquor supernatant from an offline biological reactor to an inline tank. The aeration to the offline tank is turned off and the solids are allowed to settle for 2 hours. A sump pump is then used to transfer the supernatant to the inline tank. After the supernatant has been transferred, returned solids are used to refill the offline tank. When this procedure was carried out daily for 2 weeks at the Williamsport (Pennsylvania) Sanitary Authority, scum and levels of actinomycetes were markedly reduced (Gerardi, 1986). Approximately 0.15 ML (40 000 gal) of supernatant to 0.57 ML (150 000 gal) of inline mixed liquor were used. The biological principle involved was believed to be lysis (disintegration) of actinomycetes, resulting in production of autolysins that attack the actinomycetes.

Water Sprayers. At several treatment plants, coarse water sprayers were used to reduce or dilute scum accumulation (Pipes, 1978). Other operations, however, have shown that spraying either fine or impact sprayers has no significant effect (Gerardi, 1986).

CONTROL OF VISCOUS BULKING

Methodologies for curing viscous bulking include chlorination, ozonation, nutrient addition, aeration, and reduction of F:M. Albertson (1992) reported that chlorine addition was highly effective in eliminating viscous bulking at Phoenix, Arizona. However, the effective dose for eliminating viscous bulking was not provided. On the other hand, Wanner (1994) found that chlorine addition to cure viscous bulking was not effective in a treatment plant receiving tannery wastewater.

Use of ozone to cure viscous bulking was described by van Leeuwen (1989). Ozone supplied at a dose of 1 g O_3/kg MLSS·d (1 lb/d/1000 lb) replaced the fingered zoogleal colonies with dense compact flocs.

Wanner (1994) reported a case of viscous bulking occurring in the first stage of a treatment plant receiving only concentrated tannery wastewater. The second stage was fed with a mixture of municipal wastewater and the treated effluent from the first stage. The addition of phosphorus to the first stage and reaeration of viscous sludge were not successful in controlling viscous bulking. However, when a fraction of the municipal wastewater was diverted to the first stage, the production of extracellular polymers was substantially reduced. This practice indicated that extracellular polymer production was aggravated by the lack of some nutrients–micronutrients. Similarly, Hale and Garver (1983) reported that inadequate nitrogen supply coupled with high organic loadings of carbohydrate wastewater from a winery in a pretreatment facility resulted in excessive amounts of extracellular polysaccharide production. The wastewater contained sugars, alcohols, and low-molecular-weight carboxylic acids. Viscous bulking was eliminated with the addition of ammonia as a nitrogen source in a COD-to-nitrogen ratio of as low as 149:1. Viscous bulking resulting from phosphorus deficiency in a Michigan papermill activated sludge was cured by increasing effluent orthophosphate concentration from less than 1 mg P/L to a range of 1 to 2 mg P/L (Jenkins et al., 1993).

Viscous bulking was observed to occur at the R.M. Clayton municipal WWTP when the DO concentration was low in the biological reactor coupled with high sludge blankets in the secondary clarifiers (Sezgin, 1989). Viscous extracellular polymers were observed hanging in the clarifiers near the surface. The bulking was cured by increasing both the effluent end DO concentration of the biological reactor to 5 mg/L and the RAS flowrate. Schwartz (1980) reported that, under normal operating conditions, the DO in the biological reactor was 4 mg/L and decreased to 1 mg/L during viscous bulking even though an additional blower was put in operation.

Viscous bulking was also observed in a bench-scale, completely mixed activated-sludge system when the sludge blanket in the secondary clarifier was high and the lower layers of the blanket became septic (Sezgin, 1989). Wanner (1994) reported that the growth of zoogleal organisms is always associated with the presence of readily biodegradable substrates, especially volatile fatty acids and a high concentration gradient of readily biodegradable substrates in activated-sludge processes. This would mean that zoogleal bulking may occur more frequently in activated-sludge systems with selectors and plug-flow configuration.

REFERENCES

Albertson, O.E. (1992) Control of Bulking and Foaming Organisms. In *Design and Retrofit of Wastewater Treatment Plants for Biological Nutrient Removal*. C.W. Randall et al. (Eds.), Technomic Publishing Company, Inc., Lancaster, Pa.

Albertson, O.E., and Hendricks, P. (1992) Bulking and Foaming Organism Control at Phoenix, AZ WWTP. *Water Sci. Technol.*, **26**, 3/4, 461.

American Public Health Association; American Water Works Association; and Water Environment Federation (1995) *Standard Methods for the Examination of Water and Wastewater*. 19th Ed., Washington, D.C.

Beebe, R.D., San Jose/Santa Clara Water Pollution Control Plant, San Jose, California (1983). Personal Communication.

Bendinger, B.; Rijnaarts, H.; Altendorf, K.; and Zehnder, A.J.B. (1993) Physicochemical Cell Surface and Adhesive Properties of Corynebacteria Related to the Presence and Chain Length of Mycolic Acids. *Appl. Environ. Microbiol.*, **59**, 3973.

Blackall, L.L.; Parlett, J.H.; Hayward, A.C.; Minniken, D.E.; Greenfield, P.F.; and Harbers, A.E. (1989) *Nocardia pinensis* spp. nov., an Actinomycete Found in Activated Sludge Foams in Australia. *J. Gen. Microbiol.*, **135**, 1547.

Blackall, L.L.; Tandoi, V.; and Jenkins, D. (1991) Continuous Culture Studies with *Nocardia amarae* from Activated Sludge and Their Implications for Foaming Control. *Res. J. Water Pollut. Control Fed.*, **63**, 44.

Blackall, L.L., Sevior, E.M.; Cunningham, M.A.; Seviour, R.J.; and Hugenholtz, P. (1994) *Microthrix parvicella* is a Novel, Deep Branching Member
of the Actinomycetes Subphylum. *Syst. Appl. Microbiol.*, **17**, 513.

Blackall, L.L.; Sevior, E.M.; Bradford, D.; Stratton, H.M.; Cunningham,
M.A.; Hugenholtz, P.; and Sevior, R.J. (1996) Towards Understanding the
Taxonomy of Some of the Filamentous Bacteria Causing Bulking and
Foaming in Activated Sludge Plants. *Water Sci. Technol.*, **34**, 5/6, 137.

Blackbeard, J.R.; Ekama, G.A.; and Marais, G.V.R. (1986) A Survey of
Bulking and Foaming in Activated Sludge Plants in South Africa. *Water
Pollut. Control*, **85**, 90.

Boylen, C.W., and Mulks, M.H. (1978) The Survival of Coryneform Bacteria
During Periods of Prolong Nutrient Starvation. *J. Gen. Microbiol.*, **105**,
323.

Cairns, W.L.; Cooper, D.G.; Zajic, J.E.; Wood, J.M.; and Losaric, N. (1982)
Characterization of *Nocardia amarae* as a Potential Biological Coalescing
Agent of Water-Oil Emulsions. *Appl. Environ. Microbiol.*, **43**, 362.

Cha, D.K.; Jenkins, D.; Lewis, W.P.; and Kido, W.H. (1992) Process Control
Factors Influencing *Nocardia* Populations in Activated Sludge. *Water
Environ. Res.*, **64**, 37

Chudoba, J.; Ottova, V.; and Madera, V. (1973a) Control of Activated Sludge
Filamentous Bulking. I. Effect of Hydraulic Regime or Degree of Mixing
in an Aeration Tank. *Water Res. (G.B.)*, **7**, 1163.

Chudoba, J.; Grau, P.; and Ottova, V. (1973b) Control of Activated Sludge
Bulking. II. Selection of Micro-organisms by Means of a Selector. *Water
Res. (G.B.)*, **7**, 1389.

Chun, J.; Blackall, L.L.; Kang, S.-O.; Hah, Y.C.; and Goodfellow, M. (1997)
A Proposal to Reclassify *Nocardia pinensis* Blackall et al. as *Skermania
piniformis* gen nov., comb. nov. *Int. J. Syst. Bacteriol.*, **47**, 127.

Daigger, G.T., and Roper, R.E., Jr. (1985) The Relationship Between SVI and
Activated Sludge Settling Characteristics. *J. Water Pollut. Control Fed.*, **57**,
859.

Duchene, Ph., and Pujol, R. (1991) French Research into Biological Foaming.
In *Biological Approach to Sewage Treatment Process: Current Status and
Perspectives*. P. Madoni (Ed.), Perugia, It.

Eikelboom, D.H. (1975) Filamentous Organisms Observed in Bulking
Activated Sludge. *Water Res. (G.B.)*, **9**, 365.

Eikelboom, D.H. (1977) Identification of Filamentous Organisms in Bulking
Activated Sludge. *Prog. Water Technol.*, **8**, 153.

Eikelboom, D.H., and van Buijsen, H.J.J. (1981) *Microscopic Sludge Investigation Manual.* TNO Res. Inst. Environ. Hyg., Neth.

Farquhar, G.J., and Boyle, W.C. (1972) Control of *Thiothrix* in Activated
Sludge. *J. Water Pollut. Control Fed.*, **44**, 14.

Franz, A., and Matsche, N. (1994) Investigation of a Bacteria-Enzyme
Additive to Prevent Foaming in Activated Sludge Plants. *Water Sci.
Technol.*, **29**, 7, 281.

Gasser, J.A. (1987) Control of *Nocardia* Scum in Activated Sludge by
Periodic Anoxia. *J. Water Pollut. Control Fed.*, **59**, 914.

Gerardi, M.H. (1986) Control of Actinomycetic Foam and Scum Production. *Publ. Works.*

Goi, M.; Nishimura, T.; Kuribayashi, S.; Okouchi, T.; and Murakami, T. (1995) An Experimental Study to Suppress Scum Formation Accompanying the Abnormal Growth of *Nocardia* by Adding Ozone in the Aeration Tank. *Water Sci. Technol.*, **30**, 11, 231.

Goodfellow, M.; Stainsby, F.M.; Davenport, R.; Chun, J.; and Curtis, T. (1998) Activated Sludge Foaming: The True Extent of Actinomycete Diversity. *Water Sci. Technol.*, **37**, 4/5, 511.

Goronszy, M.C. (1979) Intermittent Operation of the Extended Aeration Process for Small Systems. *J. Water Pollut. Control Fed.*, **51**, 274.

Hale, F.D., and Garver, S.R. (1983) Viscous Bulking of Activated Sludge. Paper presented at 56th Annu. Water Pollut. Control Conf., Atlanta, Ga.

Henze, M.R.; Dupont, R.; Grau, P.; and de la Sota, A. (1993) Rising Sludge in Secondary Settlers Due to Denitrification. *Water Res.* (G.B.), **27**, 231.

Hiraoka, M., and Tsumura, K. (1984) Suppression of Actinomycete Scum Production—A Case Study at Senboku Wastewater Treatment Plant, Japan. *Water Sci. Technol.*, **16**, 83.

Jenkins, D.; Neethling, J.B.; Bode, H.; and Richard, M.G. (1982) The Use of Chlorination for Control of Activated Sludge Bulking. In *Bulking of Activated Sludge: Preventative and Remedial Methods*. B. Chambers and E.J. Tomlinson (Eds.), Ellis Horwood, Ltd., Chichester, Eng.

Jenkins, D.; Richard, M.G.; and Daigger, G.T. (1993) *Manual on the Causes and Control of Activated Sludge Bulking and Foaming*. 2nd Ed., Lewis Publishers, Chelsea, Mich.

Jones, C., Utoy Creek Wastewater Treatment Plant, Atlanta, Georgia (1988). Personal Communication.

Kappeler, J.; Purtschert, I.; and Gujer, W. (1993) An Analysis of Practical Experience with Scum Formation in Full-Scale Activated Sludge Systems in Switzerland. In *Prevention and Control of Bulking Activated Sludge*. D. Jenkins (Ed.), Luigi Bazzucchi Center, Perugia, It.

Keinath, T.M. (1985) Operational Dynamics and Control of Secondary Clarifiers. *J. Water Pollut. Control Fed.*, **57**, 770.

Keinath, T.M.; Ryckman, M.D.; Dana, C.H.; and Hofer, D.A. (1977) Activated Sludge—Unified System Design and Operation. *J. Environ. Eng.*, **103**, EE5, 829.

Keller, P.J., and Cole, C.A. (1973) Hydrogen Peroxide Cures Bulking. *Water Wastes Eng.*, **10**, E4.

Khan, A.R., and Forster, C.F. (1988) Biosurfactant Production by *Rhodococcus rubra*. *Environ. Technol. Lett.*, **9**, 1349.

Khan, A.R., and Forster, C.F. (1990a) Activated Sludge Foams: An Examination into Their Stability and Their Control. *Environ. Technol.*, **11**, 1153.

Khan, A.R., and Forster, C.F. (1990b) An Investigation in the Stability of Foams Related to the Activated Sludge Process. *Enzyme Microbiol. Technol.*, **12**, 788.

Kilroy, A.C., and Gray, N.F. (1992) Pilot Plant Investigations of the Treatability of Ethylene Glycol by Activated Sludge. *Environ. Technol.*, **13**, 292.

Klatte, S.; Rainey, F.A.; and Kroppenstendt, R.M. (1994) Transfer of *Rhodococcus aichiensis* Tsukamura 1982 and *Nocardia amarae* Lechevalier and Lechevalier 1974 to the genus *Gordona* as *Gordona aichiensis* comb. nov. and *Gordona amarae* comb. nov. *Int. J. Syst. Bacteriol.*, **44**, 769.

Lechevalier, H.A. (1975) *Actinomycetes of Sewage Treatment Plants*. EPA-600/2-75-031, U.S. EPA, Cincinnati, Ohio.

Lechevalier, H. (1986) Nocardioforms. In *Bergey's Manual of Systematic Bacteriology*. Williams & Williams, Baltimore, Md., **2**.

Lechevalier, M.P., and Lechevalier, H.A. (1974) *Nocardia amarae* spp. nov., An Actinomycete Common in Foaming Activated Sludge. *Int. J. Syst. Bacteriol.*, **24**, 278.

Lee, S.-E.; Koopman, B.L.; Jenkins, D.; and Lewis, R.F. (1982) The Effect of Aeration Basin Configuration on Activated Sludge Bulking at Low Organic Loading. *Water Sci. Technol.*, **14**, 407.

Lee, S.-E.; Koopman, B.L.; Bode, H. and Jenkins, D. (1983) Evaluation of Alternative Sludge Settleability Indices. *Water Res.* (G.B.), **17**, 1421.

Lemmer, H., Bayerische Landesan stalt fur Wasserforschung, Germany (1989). Personal Communication.

Lemmer, H., and Kroppenstedt, R.M. (1984) Chemotaxonomy and Physiology of Some Actinomycetes Isolated from Scumming Activated Sludge. *Syst. Appl. Microbiol.*, **5**, 124.

Mamais, D., and Jenkins, D. (1992) The Effect of MCRT and Temperature on Enhanced Biological Phosphorus Removal. *Water Sci. Technol.*, **26**, 5/6, 955.

Marshall, K.C. (1979) Growth at Interfaces. In *Strategies of Microbial Life in Extreme Environments*. M. Shilo (Ed.), Dahlem Konferenzen, Berline, 281.

Milwaukee Mystery: Unusual Operating Problem Develops (1969). *Water Sew. Works*, **116**, 213.

Mori, T.; Saki, Y.; Honda, K.; Yano, I.; and Hashimoto, S. (1988) Stable Abnormal Foam in Activated Sludge Process Produced by *Rhodococcus* spp. with Strong Hydrophobic Property. *Environ. Technol. Lett.*, **9**, 1041.

Mori, T.; Itokazu, K.; Ishikura, Y.; Mishina, F.; Saki, Y.; and Koga, M. (1992) Evaluation of Control Strategies for Actinomycete Scum in Full-Scale Treatment Plants. *Water Sci. Technol.*, **25**, 6, 231.

Okouchi, T.; Goi, M.; Nishimura, T.; Abe, S.; and Murakami, T. (1996) *Nocardia* Scum Suppression Technology by Ozone Addition. *Water Sci. Technol.*, **34**, 3-4, 283.

Ooten, R.J., County Sanitation Districts of Orange County, California (1989). Personal Communication.

Ostrander, S.J. (1992) A Non-Conventional Solution to an Old Problem. *Oper. Forum*, **9**, 10.

Pagilla, K.R.; Jenkins, D.; and Kido, W.H. (1996) *Nocardia* Control in Activated Sludge By Classifying Selectors and Anaerobic Selectors. *Water Environ. Res.*, **68**, 235.

Palm, J.C.; Jenkins, D.; and Parker, D.S. (1980) Relationship Between Organic Loading, Dissolved Oxygen Concentration and Sludge Settleability in the Completely-Mixed Activated Sludge Process. *J. Water Pollut. Control Fed.*, **52**, 2484.

Pipes, W.O. (1978) Actinomycete Scum Production in Activated Sludge Processes. *J. Water Pollut. Control Fed.*, **50**, 628.

Pitt, P.A., and Jenkins, D. (1990) Causes and Control of *Nocardia* in Activated Sludge. *Res. J. Water Pollut Control Fed.*, **62**, 143.

Pretorius, W.A. (1987) A Conceptual Basis for Microbial Selection in Biological Waste Treatment. *Water Res.* (G.B.), **21**, 891.

Pretorius, W.A., and Laubscher, C.J.P. (1987) Control of Biological Scum in Activated Sludge Plants by Means of Selective Flotation. *Water Sci. Technol.*, **19**, 1003.

Rensik, J.H. (1974) New Approach to Preventing Bulking Sludge. *J. Water Pollut. Control Fed.*, **46**, 1888.

Rensik, J.H.; Donker, H.J.G.W.; and Ijwema, T.S.J. (1982) The Influence of Feed Pattern on Sludge Bulking. In *Bulking of Activated Sludge: Preventative and Remedial Methods.* B. Chambers and E.J. Tomlinson (Eds.), Ellis Horwood, Ltd., Chichester, Eng.

Richard, M. (1989) *Activated Sludge Microbiology.* Water Pollution Control Federation, Alexandria, Va.

Richard, M.G.; Jenkins, D.; Hao, O.; and Shimizu, G. (1982) The Isolation and Characterization of Filamentous Microorganisms from Activated Sludge Bulking. Rep. No. 81-2, Sanit. Eng. Environ. Health Res. Lab., Univ. Calif., Berkeley.

Richard, M.G.; Hao, O.; and Jenkins, D. (1985) Growth Kinetics of *Sphaerotilus* Species and Their Significance in Activated Sludge Bulking. *J. Water Pollut. Control Fed.*, **57**, 68.

Richards, T.; Nungesser, P.; and Jones, C. (1990) Solution to *Nocardia* Foaming Problems. *Res. J. Water Pollut. Control Fed.*, **62**, 915.

Rossetti, S.; Carucci, A.; and Rolle, E. (1994) Survey on the Occurrence of Filamentous Organisms in Municipal Wastewater Treatment Plants Related to Their Operating Conditions. *Water Sci. Technol.*, **29**, 305.

Scheff, G.; Salcher, O.; and Lingens, F. (1984) *Trichococcus flocculiformis* gen. nov. sp. nov.—A New Gram-Positive Bacterium Isolated from Bulking Sludge. *Appl. Microbiol. Biotechnol.*, **19**, 114.

Schuyler, R.G., Rothberg Tamburini Winsor, Inc., Denver, Colorado (1999). Personal Communication.

Schwartz, H.G.; Popowchak, T.; and Becker, K. (1980) Control of Sludge Bulking in the Brewing Industry. *J. Water Pollut. Control Fed.*, **52**, 2977.

Semon, J., Stamford Water Pollution Control Authority, Stamford, Connecticut (1985). Personal Communication.

Senghas, E., and Lingens, F. (1985) Characterisation of a New Gram Negative Filamentous Bacterium from Bulking Sludge. *Appl. Microbiol. Biotechnol.*, **21**, 118.

Seviour, E.M.; Williams, C.; DeGrey, B.; Soddell, J.A.; Sevior, R.J.; and Lindrea, K.C. (1994) Studies on Filamentous Bacteria from Australian Activated Sludge Plants. *Water Res.* (G.B.), **28**, 2335.

Sezgin, M. (1977) The Effect of Dissolved Oxygen Concentration on the Activated Sludge Process Performance. Ph.D. thesis, Dep. Civ. Eng., Univ. Calif., Berkeley.

Sezgin, M. (1982) Variation of Sludge Volume Index with Activated Sludge Characteristics. *Water Res.* (G.B.), **16**, 83.

Sezgin, M. (1989) Unpublished data. City of Atlanta, Ga.

Sezgin, M., and Karr, P.R. (1986) Control of Actinomycete Scum on Aeration Basins and Clarifiers. *J. Water Pollut. Control Fed.*, **58**, 972.

Sezgin, M.; Jenkins, D.; and Parker, D.S. (1978) A Unified Theory of Filamentous Activated Sludge Bulking. *J. Water Pollut. Control Fed.*, **50**, 362.

Sezgin, M.; Lechevalier, M.P.; and Karr, P.R. (1988) Isolation and Identification of Actinomycetes Present in Activated Sludge Scum. *Water Sci. Technol.*, **20**, 11/12, 257.

Shao, Y.J.; Wada, F.; Abkian, V.; Crosse, J.; Horenstein, B.; and Jenkins, D. (1992) Effects of MCRT on Enhanced Biological Phosphorus Removal. *Water Sci. Technol.*, **26**, 5/6, 967.

Shao, Y.J.; Starr, M.; Kaporis, K.; Kim, H.S.; and Jenkins, D. (1997) Polymer Addition as a Solution to *Nocardia* Foaming Problems. *Water Environ. Res.*, **69**, 25.

Simpson, J.R.; List, E.; and Dunbar, J.M. (1991) Bulking Sludge: A Theory and Successful Case Histories. *J. Inst. Water Environ. Manage.*, **5**, 302.

Singer, P.C.; Pipes, W.O.; and Herman, E.R. (1968) Flocculation of Bulked Activated Sludge with Polyelectrolytes. *J. Water Pollut. Control Fed.*, **40**, Part 2, R1.

Soddell, J.A.; Seviour, R.J.; Blackall, L.L.; and Hugenholtz, P. (1998) New Foam-Forming Nocardioforms Found in Activated Sludge. *Water Sci. Technol.*, **37**, 4/5, 495.

Stratton, H.; Seviour, B.; and Brooks, P. (1998) Activated Sludge Foaming: What Causes Hydrophobicity and Can It Be Manupulated to Control Foaming? *Water Sci. Technol.*, **37**, 4-5, 503.

Strom, P.F., and Jenkins, D. (1984) Identification and Significance of Filamentous Microorganisms in Activated Sludge. *J. Water Pollut. Control Fed.*, **56**, 52.

Sunairi, M.; Iwabuchi, N.; Yoshizawa, Y.; Murooka, H.; Morisaki, H.; and Nakajima, M. (1997) Cell-Surface Hydrophobicity and Scum Formation of *Rhodococcus rhodochrous* Strains with Different Colonial Morphologies. *J. Appl. Microbiol.*, **82**, 204.

Trick, I., and Lingens, F. (1984) Characterization of *Herpetosiphon* spec—A Gliding Filamentous Bacterium from Bulking Sludge. *Appl. Microbiol. Biotechnol.*, **19**, 191.

Trick, I.; Salcher, O.; and Lingens, F. (1984) Characterization of Filament Forming *Bacillus* Strains Isolated from Bulking Sludge. *Appl. Microbiol. Biotechnol.*, **19**, 120.

U.S. Environmental Protection Agency (1985) *Process Design Manual, Odor and Corrosion Control in Sanitary Sewerage Systems and Treatment Plants.* EPA-625/1-85-018, Off. Res. Dev., Cincinnati, Ohio.

van den Eynde, E.; Houtmeyers, J.; and Verachtert, H. (1982) Relation Between Substrate Feeding Pattern and Development of Filamentous Bacteria in Activated Sludge. In *Bulking of Activated Sludge: Preventative and Remedial Methods*. B. Chambers and E.J. Tomlinson (Eds.), Ellis Horwood, Ltd., Chichester, Eng.

van Leeuwen, J. (1989) Ozonation for Non-Filamentous Bulking Control in an Activated Sludge Plant Treating Fuel Synthesis Waste Water. *Water SA*, **15**, 127.

van Leeuwen, J., and Pretorius, W.A. (1988) Sludge Bulking Control with Ozone. *J. Inst. Water Environ. Manage.*, **2**, 223.

Vega-Rodriquez, B.A. (1983) Quantitative Evaluation of *Nocardia* spp. Presence in Activated Sludge. M.S. thesis, Dep. Civ. Eng., Univ. Calif., Berkeley.

Wagner, F. (1982) Study of the Causes and Prevention of Sludge Bulking in Germany. In *Bulking of Activated Sludge: Preventative and Remedial Methods*. B. Chambers and E.J. Tomlinson (Eds.), Ellis Horwood, Ltd., Chichester, Eng.

Wagner, M.; Erhart, R.; Manz, W.; Amann, R.; Lemmer, H.; Wedi, D.; and Schleifer, K.-H. (1994) Development of an rRNA-Targeted Oligonucleotide Probe Specific for the Genus *Acinetobacter* and Its Application for In Situ Monitoring in Activated Sludge. *Appl. Environ. Microbiol.*, **60**, 792.

Wanner, J. (1994) *Activated Sludge Bulking and Foaming Control*. Technomic Publishing Company, Inc., Lancaster, Pa.

Water Environment Federation (1994) *Wastewater Biology: The Life Processes*, Special Publication, Alexandria, Va.

Water Pollution Control Federation (1987) *Activated Sludge*. Manual of Practice No. OM-9, Alexandria, Va.

Wheeler, D.R., and Rule, A.M. (1980) The Role of *Nocardia* in the Foaming of Activated Sludge: Laboratory Studies. Paper presented at Ga. Water Pollut. Control Assoc. Annu. Meeting, Savannah, Ga.

White, M.J.D. (1976) Design and Control of Secondary Settlement Tanks. *Water Pollut. Control*, **75**, 459.

Williams, S.T.; Lanning, S.; and Wellington, E.M.H. (1984) Ecology of Actinomycete. In *The Biology of the Actinomycetes*. M. Goodfellow et al. (Eds.), Academic Press, New York, 109.

Wilson, T.E.; Ambrose, W.A.; and Buhr, H.O. (1984) Operating Experiences at Low Solids Residence Time. *Water Sci. Technol.*, **16**, 661.

Wong, P.K., and Chung, W.K. (1993) Control of Bacterial Foaming in Activated Sludge Process by Chlorination. *J. Environ. Sci. Health*, **A28**, 8, 1615.

Wood, D.K., and Tchobanoglous, G.T. (1974) Trace Elements in Biological Waste Treatment with Specific Reference to the Activated Sludge Process. *Proc. 29th Ind. Waste Conf., Purdue Univ. Ext. Ser. 145*, West Lafayette, Ind., 648.

Zickefoose, C.S., and Vass, R. (1984) Causes and Cures of Scum and Foam Formation in Treatment Plants. Brown & Caldwell Rep., presented at Calif. Water Pollut. Control Assoc. Training Conf., San Bernadino, Calif.

Chapter 9
Indicator Bacteria

CHARACTERIZATION

Coliform bacteria, especially the fecal coliforms, are natural, typically harmless inhabitants of the intestines of all warm-blooded animals, including humans. Coliforms coexist in fecal material with pathogens (i.e., disease-causing organisms) such as certain bacteria, viruses, protozoa, and helminths (worms). Although coliform bacteria are most abundant in fecal material, they may also be found in soil and on vegetable matter.

Coliforms are highly concentrated in wastewater containing fecal matter and typically sparse in unpolluted surface waters. Because of this correlation between coliforms and wastewater, the presence of coliform bacteria in water is considered an indication of wastewater contamination. Fecal-contaminated water can transmit bacterial diseases such as typhoid fever, dysentery, gastroenteritis, and cholera; viral diseases such as hepatitis and polio; and intestinal parasites such as *Giardia*, *Cryptosporidium*, and *Entamoeba*.

TOTAL COLIFORMS. Total coliform bacteria (TC) include the fecal coliforms (FC) and a wide variety of other species. The TC are Gram negative, aerobic and facultative anaerobic, nonspore forming, rod-shaped bacteria. They are typically associated with fecal material, but some species thrive on certain types of vegetation, in soils, and in some industrial processes. Species of *Escherichia*, *Enterobacter*, *Klebsiella*, and *Citrobacter* are important members of the TC group. The TC group includes harmless, pathogenic, and opportunistically pathogenic species.

FECAL COLIFORMS. Fecal coliforms are a subgroup of the TC group. They are thermotolerant and thus are capable of living at elevated or warm-blooded temperatures. The most abundant species in the FC group are *Escherichia coli* and *Klebsiella* spp. Typically, *E. coli* comprise approximately 95% of the FC recovered from most waters. *Klebsiella* may dominate in waters receiving certain industrial discharges, such as those from paper mills and food processing plants. Typical FC densities (concentrations) are given in Table 9.1.

ESCHERICHIA COLI. *Escherichia coli* is a natural inhabitant only in the intestines of warm-blooded animals. Its presence, therefore, demonstrates fecal contamination and the possible presence of intestinal or enteric pathogens. Although most strains of *E. coli* are harmless, there are several pathogenic types including the type 0157:H7 that causes a sometimes fatal bloody diarrhea.

Table 9.1   Typical concentrations of fecal coliform bacteria.

Source	Organisms/100 mL undisinfected sample		
	Average	Minimum	Maximum
Raw wastewater	8×10^6	3.4×10^5	4.9×10^7
Secondary effluent	5×10^5	1.1×10^4	1.59×10^6
Disinfected effluent	100	0	2400
Human feces	5×10^{10}	1×10^9	1×10^{12}
Unstabilized sludge	1×10^9		
Anaerobically digested solids	3×10^4–6×10^6		
Mixing zone for secondary effluent	1×10^4		

FECAL STREPTOCOCCI. The fecal streptococci (FS) include the entero-coccus group and several species (*Streptococcus bovis*, *Streptococcus equinus*, *Streptococcus avium*, and *Enterococcus faecalis liquefaciens*) that are associated with nonhuman warm-blooded animal wastes and vegetation. High concentrations of FS indicate contamination with animal feces and the possible presence of pathogenic organisms. An identification and enumeration of the FS species present can be used to determine the source or sources of contamination.

ENTEROCOCCI. Enterococci are a subgroup of the fecal streptococci. They are enteric bacteria of humans and other warm-blooded animals and consist primarily of *Enterococcus faecalis zymogenes* (formerly *Streptococcus faecalis zymogenes*) and *Enterococcus faecium* (formerly *Streptococcus faecium*). Approximately 80% of the human FS are *E. faecalis zymogenes*. Additional characterization of the recovered enterococci is necessary to determine those that are *E. faecalis liquefaciens*. This subspecies is not restricted to the intestines of warm-blooded animals and has been found associated with vegetation, insects, and certain soils. It typically predominates at densities less than 100 organisms/100 mL. Enterococci densities greater than 100 organisms/100 mL indicate contamination of warm-blooded animal (and possibly human) origin with the possible presence of pathogens.

USE OF BACTERIA AS INDICATOR ORGANISMS

The risk associated with the presence of pathogenic organisms in any fecal-contaminated water is assessed indirectly by determining the concentration of an indicator organism. This provides a measure of the magnitude of contamination present.

National and state water quality standards (e.g., the 200 FC/100 mL limit for full body contact recreational waters) carry an associated health risk based on the probable presence of pathogens. Current national and state regulations use TC, FC, *E. coli*, and enterococci as indicator organisms.

Fecal coliforms have several of the qualities of a good indicator organism: they are closely associated with fecal contamination, relatively easy to identify and enumerate, and are typically present in large numbers when fecal contamination is present. Fecal coliforms are the principal indicator organisms used for evaluating the microbiological contamination of recreational waters.

Other indicator organisms, including *E. coli*, FS, and enterococci, are routinely used in certain areas to assess water quality. *Escherichia coli* and FC are the recommended indicators for fresh waters. Enterococci and FS are recommended for both fresh and salt water because of their correlation with the outbreak of swimming-related diseases (especially gastroenteritis).

Because of their survival behavior in the environment, it has been suggested that FS and enterococci are better indicators of viral contamination than coliforms.

INDICATOR SENSITIVITY. Total coliforms are by definition more abundant than FC by approximately 3 to 5 times, although this ratio varies widely. Because of their greater abundance, TC provide a more representative measure of wastewater treatment effectiveness than do FC by permitting the detection of even slight amounts of contamination. This sensitivity is useful when evaluating clean water, such as shellfishing areas and drinking water. Natural, nonwastewater sources of TC, such as animal sources and decaying vegetation, may produce a positive test result causing a sample to be mistakenly classified as contaminated by wastewater. This possibility exists to a lesser extent with the FC (approximately 95% of which are *E. coli*), which have few nonfecal sources.

The fecal coliforms are more closely associated with fecal discharges than are TC. They are less subject to regrowth, except in sediments, and are less common in unpolluted areas than TC. Their presence in surface waters in excess of the 200 FC/100 mL limit indicates an unacceptable level of fecal contamination.

Indicator organism ratios can help to identify sources of pollution. Fecal coliform-to-fecal streptococci ratios greater than 4.0 indicate fecal pollution from human sources, ratios less than 0.7 indicate nonhuman sources and ratios between indicate a mixture of sources. Because of the rapid die-off of some FS outside of the animal host, these ratios are valid only within 24 hours following their discharge.

INDICATOR LIMITATIONS. Although indicators are useful in assessing the overall microbiological contamination of natural waters, no indicator species or group is perfect. Typically, a high concentration of an indicator suggests high concentrations of pathogens. Exceptions to this are that pathogens may be abundant when indicators are not, e.g., viruses, and vice versa. These exceptions may result from natural effects, water pollution, or measurement interferences, and can cause water to be incorrectly classified.

A false or erroneous classification as uncontaminated can result when coliform concentrations are low but pathogens are high. This error can occur because some pathogens die off (or disappear, see below) slowly in water, but coliforms die off rapidly. This phenomenon makes coliforms a valid indicator only for recently contaminated water. Other conditions that may alter expected relationships between coliform and pathogen densities include the ability of some pathogens (e.g., *Vibrio* spp.) to multiply in natural waters, occurrence of epidemics within a population resulting in the release of greater numbers of pathogens when coliform densities remain typical, or laboratory testing difficulties that give erroneous coliform test results. Underestimated coliform concentrations may result from the interfering growth of noncoliform species, the presence of heavy metals, or chlorine. Suspended solids may cause abnormal colony formation, blurring of colonies, and clogged membrane filters.

Proper training, attention to detail, and close adherence to standard procedures are essential to obtaining reliable results. Although not perfect, coliform bacteria typically provide a good assessment of recent local wastewater pollution and are useful for legal and decision-making purposes.

INDICATOR DIE-OFF (DISAPPEARANCE) RATES. Coliforms have a half-life of approximately 15 hours and die off rapidly in water. Few coliforms will be recovered from the water column after more than 3 days without additional inputs (i.e., nutrients and new contamination) from outside sources. Although the term die-off is used, it should be noted that indicator organisms may disappear from the water column because of sedimentation, predation, dilution, and death. Because of this rapid die-off, coliforms tend to be restricted to the vicinity of the discharge source. In contrast, some FS, other resistant bacteria, viruses, protozoa, and pathogens may survive much longer than coliforms in receiving water. Pathogens may be present in water that no longer contains recoverable coliforms and may travel downstream to endanger areas distant from the original discharge point. Fecal coliforms have been shown to survive for extended periods of time in freshwater sediments, estuarine sediments, marine sediments, and storm drain sediments.

Aquatic sediments are considered to be a reservoir for microorganisms including TC, FC, FS, and enterococci. Indeed, bacterial concentrations are much greater in sediments, than in the overlying waters. Natural and anthropogenic events can resuspend these bacteria-laden sediments, greatly increasing indicator and pathogen concentrations in the overlying waters. Resuspended bacteria can obscure the source of the contamination; that is, whether it is from recent discharge, bypass, overflow, leak, or spill, or from an event more distant in time. Regardless of the immediate source, high indicator concentrations are a potential public health concern because of the possible presence of pathogens; for example, enteropathogenic *E. coli* and opportunistically pathogenic *Klebsiella pneumoniae* (members of the FC group). Ultraviolet radiation (sunlight) and elevated temperatures both serve to increase the die-off rate of indicator bacteria. Other factors such as salinity, nutrients, heavy metal concentrations, sedimentation, competition, antagonism, and predation also affect indicator bacteria survival.

REGULATIONS AFFECTING INDICATOR ORGANISMS.
Microbiological water quality standards and effluent limitations assume a quantitative relationship between health risk and concentration of indicator organisms. Regulations are set after considering health risks, the likelihood of compliance, economics, aesthetics, and other factors.

Effluent Limits. All wastewater treatment plant discharges in the United States are subject to National or State Pollution Discharge Elimination System (NPDES or SPDES) permits. Fecal coliform discharge limits are based on water quality criteria set by the states and approved by the U.S. Environmental Protection Agency (U.S. EPA). Hence, permit requirements may vary by state and region.

Receiving Water Criteria. Acceptable (safe) levels of FC in receiving water bodies depend on water body use. Only low FC levels are allowed in swimming areas but higher levels may be acceptable elsewhere. Each state is required by U.S. EPA to specify water uses to be achieved and protected. These uses include public water supplies; protection and propagation of fish, shellfish, and wildlife; recreation in and on water; and agricultural, industrial, and other purposes including navigation. In the recommended U.S. EPA water quality criteria, the indicator organisms to use, the method of measurement (including the duration and frequency of sampling), and certain statistics for analyzing the data are given. The individual states are responsible for developing legal standards from the criteria. These standards tend to vary among states.

SECTION 503 REGULATIONS AND INDICATOR BACTERIA.
Currently, U.S. EPA 40 CFR Part 503 regulations designate FC and *Salmonella* spp. as the indicator organisms to be used in compost testing. The protocols are ambiguous regarding such matters as holding time, sample number and location, sample type, and measurement. For example, holding time ranges from 6 hours to 2 weeks. Indicator bacteria must be recovered within 6 hours although viruses may be recovered up to the 2-week limit. The time period used for sampling (e.g., within 24 hours, 5 days, or 30 days), number of samples, sample location (e.g., center or periphery of the pile), and sample type (grab or composite) all affect the measured indicator bacteria concentrations. Note that the regulations specify measurement on a dry weight basis; however, drying the sample before testing will significantly reduce the number of recoverable indicator bacteria and yield inaccurate results.

PUBLIC HEALTH IMPLICATIONS OF INDICATOR BACTERIA

Most indicator bacteria are harmless to humans; however, there are several of public health significance. The TC group includes pathogenic *E. coli* and opportunistically pathogenic *Klebsiella*, *Citrobacter*, and *Enterobacter* species. The FC group includes pathogenic *E. coli* and opportunistically pathogenic *K. pneumoniae*. *Escherichia coli* include several pathogenic strains including type 0157:H7. Approximately 2 to 8% of *E. coli* recovered from water have been found to be enteropathogenic. Species included in the FS and enterococci (i.e., Lancefield's serological group D *Streptococus*: *E. faecalis, E. faecium, Streptococcus avium,* and *S. bovis*) can cause gastroenteritis with diarrhea, cramps, nausea, vomiting, fever, chills, and dizziness within 2 to 36 hours. Although typically associated with contaminated food, waterborne transmission could occur if an infectious dose is ingested (approximately 10^7 organisms/100 mL) in water or wastewater. Combined and sanitary sewer overflows, wastewater treatment plant bypasses, wastewater before disinfection, and agricultural point and nonpoint flows (e.g., animal feed lot drainage, pasture land runoff) could provide such infectious doses. The presence of FS and enterococci indicates fecal contamination that may include pathogens and infectious doses of pathogenic streptococcal species.

Pathogenic strains of *E. coli* cause moderate to severe gastroenteritis with watery or bloody diarrhea, accompanied by nausea, abdominal cramps, and vomiting. A small percentage of cases are fatal. *Klebsiella pneumoniae* can cause a variety of illnesses including sinusitis, pharyngitis, peritonitis, and endocarditis. Its most typical manifestation is as a serious and sometimes fatal form of pneumonia. *Citrobacter* and *Enterobacter* spp. may also cause gastroenteritis and urinary tract infections. Transmission is by ingestion of contaminated water or food. Contamination can occur directly (fecal material in the water or food) or indirectly through poor hygienic practices. Those most at risk are the young, the elderly, those in poor health, and the immuno-compromised (i.e., those taking immunnosuppressant drugs, undergoing chemotherapy, or those with acquired immune deficiency syndrome).

Health risks can be reduced by following the recommended safe work and laboratory practices described in various publications issued by Occupational Safety and Health Administration, U.S. EPA, and Water Environment Federation. The importance of good personal hygiene and safe work practices, including the use of protective outer wear and the proper acquisition, handling, and processing of samples, cannot be overemphasized.

COMPONENTS OF A MONITORING PROGRAM

To demonstrate compliance with water quality regulations, indicator bacteria concentrations must be measured in both effluents and receiving waters. A good monitoring program must include multiple samplings to determine average and maximum effluent bacterial levels. The location, method, size of the sample collected, time elapsed between collection and analysis, method of analysis, and calculation of results are all critical to the monitoring program. An error in any one factor can cause large errors in the reported test findings, resulting in mistaken permit violations.

Accurate measurement of coliforms is difficult to achieve and extreme care is required in following the established procedures to prevent significant errors in the results. Procedures for measuring the indicator organisms discussed in this chapter are described in *Standard Methods for the Examination of Water and Wastewater* (APHA et al., 1995) and *Microbiological Methods for Monitoring the Environment* (U.S. EPA, 1978).

SAMPLING. Sampling Preparation. Appropriate containers for FC samples may be of glass or autoclavable, nontoxic polypropylene. They are typically 125 or 250 mL (4 or 8 oz) in volume, have secure stoppers or screw caps, and

---

are free of chipped, scratched, or etched surfaces. Before use, the bottles must be carefully cleaned and sterilized. Before sterilization, dechlorinating or chelating agents (for water samples high in chlorine or heavy metals) are added to the bottle if necessary. Disposable presterilized containers are also available for sample collection.

Sample Locations. Effluent samples should be representative of the material discharged to the receiving waters. The precise sampling location is important and is listed in the NPDES permit. Questions about sampling should be discussed with the appropriate regulatory agency.

In-stream sampling can also provide useful results and should include sampling sites upstream, downstream, and at the point of effluent discharge. Samples should be collected at times showing the variety of conditions possible (high flow, low flow, morning, afternoon, night, wet weather, and dry weather). It is important to identify causes of worst-case conditions and to include samplings during these conditions. Some procedures specify sampling only during worst-case conditions. Special studies, such as hourly measurements over several 24-hour periods, can supply useful information about the changes in coliform and other indicator organism concentrations at a specific treatment plant. A good sampling strategy will provide randomly selected representative samples during each sampling event.

Sampling Schedule. The timing of sample collection can have a significant effect on analytical results. Factors such as temperature and sunlight may affect indicator bacteria survival and can cause differing results between early morning and midafternoon samples. Practical constraints, such as the availability of an analyst to process the samples immediately after collection, can affect sample holding times and analytical results, and should be considered when designing a sampling program. A safe procedure for choosing sample times is to randomly select sampling times (preferably using random number tables), within practical constraints, to prevent biased results. The randomly selected representative samples should be taken during worst-case conditions, such as the early morning discharge.

Collection Procedures. Proper sampling can have a greater effect on analytical results than any other procedure. Obtaining a representative and uncontaminated sample for bacterial analysis requires special care and attention to detail. If raw wastewater, for example, contains FC at a concentration of 80 000 organisms/mL, it is easy to calculate that even a tiny drop of raw wastewater accidentally deposited in a 250-mL sample bottle of otherwise well-disinfected effluent will cause the analytical result to be far in excess of the permitted amount. For example, a small drop of raw wastewater (0.1 mL) will add 8000 FC, which is equivalent to 3200 additional organisms/100 mL. This error would most likely result in a report of a discharge permit violation.

The following procedures can help ensure uncontaminated samples:

- Use extra care when collecting samples.
- Be sure containers have been sterilized and contain dechlorinating or chelating agents as needed.

- Do not touch the inside of the bottle or cap.
- Do not allow any foreign material to enter the bottle or cap.
- Never place the cap where it might become contaminated.
- When sampling from a tap, avoid leaky fixtures that may contaminate the sample and flame any taps used.
- Always allow a piping system to run long enough to completely flush itself with at least 3 to 5 times its volume to ensure a fresh sample of living coliforms that is representative of the effluent.

Labeling Requirements. Complete and careful labeling of the sample bottles using a permanent, waterproof, nonsmearing marker is important. The label should contain the type, location, date, and time of the sample, and the initials of the person who collected the sample. Any additional information, such as the type and amount of dechlorinating or chelating agent added and the chlorine residual at the time of sampling, should also be included.

Because a laboratory analyst may receive a variety of samples from several different plants, a clearly written, complete label will help prevent costly laboratory errors, minimize requirements for resampling, and reduce the chance for erroneous monitoring reports. In addition to labeling, properly completed chain-of-custody forms should accompany any samples that are subject to legal scrutiny.

Sample Storage. Immediately after collection, samples should be placed on ice in an insulator or a refrigerator at 4 °C (39 °F). Care should be taken that sample bottles are not totally immersed in water during transit or storage because immersion may increase the chance of contamination. Laboratory analysis should be started as soon after collection as possible because bacteria in wastewater samples are especially susceptible to changes in concentration. Holding times greater than 6 hours are likely to provide unrepresentative results and will invalidate any results used for legal purposes.

TESTING METHODS. Three techniques that are widely used to enumerate FC in water and wastewater are the direct and the two-step membrane filter (MF) technique and the multitube fermentation or most probable number (MPN) technique. They are described in *Standard Methods* (APHA et al., 1995) and *Microbiological Methods for Monitoring the Environment* (U.S. EPA, 1978).

Membrane Filter Technique. The direct MF technique is typically favored for effluent monitoring because of its relative simplicity and minimal requirements for space, equipment, and supplies. The MF procedure requires preparation of several dilutions from the sample. Fecal coliforms are measured in each diluted sample by passing the sample through a membrane filter that retains the bacterial cells on its surface. The filter is then placed either on a pad saturated with bacterial growth medium (m-FC broth) or on m-FC agar in a petri dish that has been covered and turned upside down (inverted), then incubated.

Inversion prevents any condensation from disturbing the forming colonies and causing a false high count. High incubation temperatures and selective media prevent most nonfecal coliform bacteria from growing. Precise control of the incubation temperature is essential and is best obtained using a water bath or solid incubator. One-half degree of excess temperature will cause the FC to die off.

After 24 hours incubation, each of the FC bacteria that were originally captured on the filter will have multiplied, forming a visible, circular, blue colony. Good colony formation is necessary to give valid results and is possible only for samples that have been properly diluted and lack interfering substances. Some of these colonies are then tested to confirm their identity as FC.

Two-Step Membrane Filter Technique. The two-step MF technique is similar to the direct MF technique except that additional media are used and the incubation is started at a lower temperature to allow injured, yet viable FC (as may occur in chlorinated effluents) to recover. Wastewater containing residual chlorine, heavy metals, phenols, caustics, or acids may be candidates for the two-step incubation MF test. Neither of the MF techniques is feasible for wastewater containing heavy loads of suspended solids because excessive solids cause confluence (growing together) of colonies and other atypical or abnormal colony growth.

Comparison of Test Methods. The MPN procedure typically provides optimal conditions for FC growth, even when the sample contains toxics and solids. The procedure is less affected by toxicity, turbidity, salts, sediments, heavy metals, and chlorine than the MF procedure is because of dilution effects, better growth potential, and built-in statistical bias. As such, the MPN procedure yields a conservative measure that should be used periodically to evaluate the MF tests. The MPN method is typically used by U.S. EPA in enforcement situations.

Comparisons of the MPN and MF procedures have shown that under appropriate conditions, the two methods can provide equivalent results. However, turbidity, sediments, and saltwater are particularly deleterious to MF tests and often require the use of the MPN method for correct results. The primary drawbacks of the MPN test are that it requires more equipment and training to perform and has a longer processing time than does the MF test.

QUALITY CONTROL AND QUALITY ASSURANCE. The validity of the coliform measurements should be confirmed using a variety of quality control tests. These tests include measurements of the reproducibility and accuracy of the analytic results. A minimal quality control testing program should include duplicate MF counts, analysis of split samples, analysis of blank samples, thorough recordkeeping, and ongoing audits of results. All procedures, from sample collection to analysis, require quality control.

Duplicate Counts. Duplicate counts of a single series of MF plates are performed to ensure that results are reproducible between analysts and for a single analyst. New data sheets should be used for the second count to allow honest appraisal of the reproducibility of the first count. Duplicate counts are the most useful during training of new analysts, but provide good checks even for experienced analysts. When more than one analyst is present, it is good practice to periodically require all analysts to count the same plates, and then to compare the results obtained. Note that a small amount of variability in the counts is expected.

Split Samples. Duplicate analyses of a single split sample provide two measurements of a single original sample and should be performed on approximately 10% of the samples. This assurance procedure is necessary to demonstrate reproducibility of analyses and will quickly reveal any problems resulting from inadequate or improper mixing and contamination.

Blank Samples. Blank analyses attempt to measure coliform levels in dilution water. The analyses are typically performed daily and help to reveal contamination problems. If growth (fecal or nonfecal) is found in the blanks, the day's measurements should be considered invalid. If it is a frequent problem, the source of the problem (dilution water, media handling methods, or other procedures), should be isolated and corrected.

Environmental Blanks. Environmental blanks are used to monitor aerosol contaminants that deposit on a prepared, sterile, culture plate. The plate is left open on a laboratory bench, typically for approximately 15 minutes and then is incubated to demonstrate that samples are not contaminated by atmospheric fallout.

Recordkeeping. The results of all quality control procedures should be recorded in a log book. They can also be graphed on control charts. When problems are identified, they should be corrected as rapidly as possible. In some cases, it may be advisable to add a note to regulatory reports explaining any unusual quality control results. Such quality control documentation can be useful for regulatory purposes and is also helpful during analysis of past data.

Data Analysis. The measured concentrations of FC are typically considered statistically when compliance with regulations is determined and when accuracy and precision of analysis is assessed. Typical calculations include standard deviation, ranges, control limits, and frequency distributions. These are described in basic statistics and quality assurance–quality control references. Computer programs are also available for making these calculations.

REFERENCES

American Public Health Association; American Water Works Association; and Water Environment Federation (1995) *Standard Methods for the Examination of Water and Wastewater.* 19th Ed., Washington, D.C.

U.S. Environmental Protection Agency (1978) *Microbiological Methods for Monitoring the Environment: Water and Wastes.* EPA-600/8-78-017, Cincinnati, Ohio.

SUGGESTED READINGS

Bitton, G. (1994) *Wastewater Microbiology.* Wiley-Liss, Inc., New York.

Cabelli, V.J. (1983) *Health Effects Criteria for Marine Recreational Waters.* EPA-600/1-80-031, U.S. EPA, Research Triangle Park, N.C.

Dufour, A., and Ballentine, R. (1986) *Ambient Water Quality Criteria for Bacteria—1986.* EPA-440/5-84-002, U.S. EPA, Washington, D.C.

Gannon, J.J.; Busse, M.K.; and Schillenger, J.E. (1983) Fecal Coliform Disappearence in a River Impoundment. *Water Res.* (G.B.), **17**, 1595.

James, A., and Evison, L. (Eds.) (1979) *Biological Indicators of Water Quality.* Wiley & Sons, New York.

Koren, H., and Bisesi, M. (1996) *Handbook of Environmental Health and Safety: Principles and Practices, Volumes I and II.* 3rd Ed., CRC Press, Inc., Boca Raton, Fla.

Marino, R.P., and Gannon, J.J. (1989) Survival of Fecal Coliforms and Fecal Streptococci in Storm Drain Sediment. *Water Res.* (G.B.), **25**, 1089.

Meiscier, J.J., and Cabelli, V.J. (1982) Enterococcus and Other Microbial Indicators of Municipal Sewage Effluents. *J. Water Pollut. Control Fed.*, **54**, 1599.

Millipore Corporation (1984) *Water Microbiology: Laboratory and Field Procedures.* Bedford, Mass.

Mitchell, R. (Ed.) (1979) *Water Pollution Microbiology.* Wiley & Sons, New York.

Pipes, W.O. (Ed.) (1982) *Bacterial Indicators of Pollution.* CRC Press, Inc., Boca Raton, Fla.

U.S. Environmental Protection Agency (1985) *Test Methods for Escherichia coli and Enterococci in Water by the Membrane Filter Procedure.* EPA-600/4-85-076, Cincinnati, Ohio.

U.S. Food and Drug Administration (1997) *Foodborne Pathogenic Microorganisms and Natural Toxins Handbook.* 2nd Ed., Cent. Food Safety Appl. Nutrition, Washington, D.C.

Water Environment Federation (1991) *Biological Hazards at Wastewater Treatment Facilities.* Special Publication, Alexandria, Va.

Chapter 10
Wastewater Pathogens

CHARACTERIZATION

Fecal bacteria in raw wastewater are predominately those derived from the feces and infected urine of humans plus an intermittent addition of excrement from dogs, cats, and wildlife in stormwater runoff entering combined sewer systems. Soil organisms are also a significant component of the wastewater flora, entering through deteriorating collecting networks, street flushings, stormwater runoff, automatic car washing operations, and from the processing of garden produce in markets, homes, and restaurants. Although wastewater collection systems have decreased the public health risk of contact with fecal pathogens in urban centers, this practice only transports the collected wastes to a location where treatment may be applied before release to a water body.

The density of viable organisms in wastewater ranges from approximately 1 to 300 mil. organisms/100 mL. The great variation in these microbial populations is related to many factors, including populations served by wastewater collection systems, effects of stormwater runoff, physical and chemical characteristics of the wastewater mixture, degree of diffusion of the microfecal pellets, retention time in the collection system, season of the year, and the hour of microbiological sampling. The discharge of certain industrial wastes, including phenols, acids, alkaline substances, and heavy metal ions, gives rise to low bacterial counts whereas any introduction of wastes rich in organic nutrients may promote rapid multiplication of substantial portions of the bacterial population.

The factor that makes raw wastewater so dangerous to public health is the greatest variable of all: the transient nature and virulence of various pathogens that are only a small fraction of the total microbial content of municipal wastewater. Pathogens are defined as disease-producing microorganisms that are found among microorganisms that are characterized as bacteria, viruses, fungi, protozoans, and parasitic worms. Some pathogens (e.g., *Salmonella typhi* and *Vibrio cholera*) are extremely invasive at a relatively low infective doses upon contact with the body, regardless of age or state of vigor. Exposure to these organisms guarantees catching the disease. In addition to these obvious pathogens, there are also other organisms present in raw wastewater that may infect wastewater treatment plant (WWTP) workers, given the opportunity to overcome a healthy body's defenses, because of massive dose exposures that can be present under the circumstances found in treatment plants. These organisms (e.g., *Legionella, Mycobacterium, Aspergillus,* and *Candida*) are known as opportunistic pathogens partly because of their ability to compromise the individual's natural resistance to infection or be invasive among patients in the hospital.

TYPES OF DISEASE EXPOSURE. Enteric bacteria, viruses, protozoans, fungi, and small worms that can colonize the intestinal tract or other organs of the body have the potential to be a health threat. Examples of enteric diseases range from gastroenteritis and hepatitis to cholera. In humans, the most common symptoms are abdominal cramps and diarrhea. Fever, nausea, and vomiting are frequently found as additional symptoms. Although their potential as a health risk to workers at WWTPs has not yet been clearly demonstrated, it is prudent to avoid situations that bring these organisms in repeated contact with treatment plant personnel.

PATHWAYS FOR PATHOGEN SPREAD. Human contact with pathogens may take several different routes: ingestion of contaminated water or exposure to polluted recreational water, consumption of salad crops grown in poor-quality irrigation water, eating of raw or partially cooked shellfish harvested from polluted water, eating of food prepared under poor sanitation practices, and person-to-person contacts with individuals practicing poor hygienic habits. In all of these situations, poor-quality water is involved and the primary sources of pathogen involvement are wastewater discharges to surface water and aquifers and stormwater flushing of fecal discharges from farm animals and wildlife into receiving streams.

RISK TO PERSONNEL. Wastewater treatment, solids-handling facilities, land applications of solids, and compost operations have been considered potential sources of airborne pathogens. Many wastewater treatment processes produce large quantities of aerosols that might contain pathogens. Under suitable conditions these aerosols and particulates can carry pathogens considerable distances. Both respiratory and enteric bacteria and viruses have been implicated by their ability to disperse in this fashion.

The proximity of a WWTP to a residential area is of concern because of the potential for exposure of surrounding populations to aerosols emitted from treatment plant processes. The possibility of exposure to greater densities of aerosols from a wastewater environment for nearby residents and treatment plant workers is often greater during the night than during the day. The greater densities have been attributed to increased frequency of atmospheric inversions, wind movement, and humidity. The application of effluent and sludges to agricultural lands has the potential of spreading microbial contamination to food crops, surface water, and groundwater sources. This is confirmed by a study of application of wastewater solids to agricultural land that indicated the prevalence of enteric bacterial pathogens. Furthermore, the study also found an increased immune response for individuals working in solids-handling operations, suggesting exposure brought greater potential health risks. Wastewater conveyance systems and treatment processes expose wastewater personnel to many pathogenic organisms and potential occupational health hazards. However, for most workers, the risk of developing a disease is low, provided employees have access to clean uniforms, shower rooms, clean restrooms with adequate hygienic supplies, and prohibitions against eating in the area of plant treatment operations. Additional protection from some diseases can be obtained through standard immunizations, which must include periodic boosters for typhoid and tetanus. An active immunization is available for hepatitis A but requires a booster shot 6 to 12 months later to sustain its effectiveness. Furthermore, it is recommended that retesting be done each year to ensure that the immunization is still active; if not, another booster shot must be given to those individuals that have lost their protective immunity.

PATHOGEN OCCURRENCES IN WASTEWATER AND SLUDGE

The density and variety of human pathogens in raw wastewater is related to the population served by the wastewater collection system, by seasonal patterns for certain diseases, and the extent of community infections at a given time. The absence of any single pathogen should not be interpreted as signifying the absence of other pathogens in wastewater. These agents originate in feces of animals (humans, pets, and wildlife) that are ill or are carriers of some pathogens but show no symptoms of sickness. Not all animals, including humans, will be continually shedding pathogenic organisms, but a low percentage of some warm-blooded animals are carriers (reservoirs) for pathogens without showing any outward symptoms of illness.

Much of the knowledge about human pathogens in wastewater has been learned from studies involving *Salmonella*, largely because it is easy to detect. The incidence of human salmonellosis, although seasonal, is fairly low and a minimum population is necessary before isolation of *Salmonella* strains in wastewater is possible. The average number of individuals excreting *Salmonella* at a given time is not known for certain, however it has been reported to vary from 1% in the United States and Great Britain to 3% in Australia and 3.9% in Ceylon (Geldreich and Van Donsel, 1970). This suggests that *Salmonella* would not be expected from domestic wastewater from a small community most of the time, although significant levels of fecal coliforms and possible other pathogens might be present at any time.

It has been hypothesized that a wastewater collection network of 50 to 100 homes is the minimum size required before there is a chance of the successful detection of some *Salmonella* strain. This hypothesis was confirmed by another study that reported *Salmonella* strains were regularly found in the collection system of a residential area of 4000 people (Harvey, et al., 1969; Callaghan and Brodie, 1969; and Geldreich, 1996).

BACTERIAL AGENTS. There is a wide variety of bacterial pathogens that may be found in wastewater and sludges. Also variable is the number of bacteria required to cause infection, ranging from a few cells for the typhoid fever bacillus and *Vibrio cholera* to approximately 100 cells for *E. coli* 0157:H7 and thousands for some of the *Salmonella typhimurium* strains involved in waterborne outbreaks. Other variables include frequency of isolation from a given animal and variations in occurrence from season to season and year to year. For these reasons, any fecal contamination must be considered to contain some pathogenic agent and therefore pose a public health risk.

Campylobacter. In humans, this pathogenic bacterial disease typically occurs as diarrhea (sometimes with blood in the stool), abdominal pain, fever, nausea, and sometimes vomiting. The pathogen (*Campylobacter jejuni* or

Campylobacter fetus) is transmitted by the fecal–oral route through contaminated water sources or raw (unpasteurized) milk. Sources of the agent may be from the feces of farm animals (chicken, turkeys, pigs, cattle, and sheep). Wildlife (beaver, deer, coyote, and gulls) that are permanent residents on the watershed may also serve as reservoirs for *Campylobacter* and contaminate remote mountain streams that may be used as a vacationer's drinking water source.

Enteropathogenic *Escherichia coli*. Most *E. coli* bacterial strains are nonpathogenic and a normal inhabitant of the intestinal track of all warm-blooded animals. However, there are several serotypes that can infect the intestinal track to cause a gastroenteritis characterized by a profuse watery diarrhea. Other pathogenic strains such as *E.coli* 0157:H7 cause a serious hemorrhagic colitis (bloody, watery diarrhea with mucus discharge, nausea, prostration, dehydration) and even death if not quickly treated. Source of these agents is most often associated with undercooked hamburger meat contaminated with bovine feces but occasionally the pathway is through drinking contaminated water. The infective dose for this pathogenic E. coli is estimated to range from 10 to 100 organisms per 100 mL with infectivity the most severe among infants, senior citizens, and patients in nursing homes (Swerdlow, et al., 1992).

Leptospira. These bacteria include a group of coil-shaped, actively motile organisms, some of which are pathogenic. They are discharged in the urine of a wide variety of infected wildlife, including rodents, and farm animals. Body contact through skin abrasions or mucus membranes provides opportunities to invade the blood stream and produce acute infections involving the kidneys, liver, and central nervous system. Many of the more than 100 serotypes of *Leptospira* have been isolated from human cases and are primarily associated with occupational exposures; for example, agricultural activities involving farm animals, irrigation of farm fields, production of rice and sugar cane, processing of fish and poultry, and exposure to stagnant water drainage in mines and other rat infested surroundings. Summertime bathing in farm ponds and slow-flowing streams around livestock pastures are also a significant source of contact with this pathogen. Although *Leptospira* have a short survival rate (12 to 14 hours) in raw wastewater because of competition from other organisms in the high-density microbial flora, soil contaminated with urine from infected animals retains viable leptospires for at least 15 days, suggesting there is a potential risk to operators working around waste stabilization lagoons.

Salmonella. Numerous serotypes (approximately 2000 different types) of *Salmonella* are pathogenic to man and other animals. Human enteric diseases from these organisms may range from self-limiting gastroenteritis caused *Salmonella typhimurium* infections with mild to severe symptoms of short duration, to severe gastroenteritis caused by *S. typhi* that is debilitating and potentially fatal. These pathogens are the most common cause of diarrhea in the United States. The total number of *Salmonella* strains known to be pathogenic to man exceeds several hundred and their frequency of isolation varies from year to year and country to country. The source of infection is fecal contamination and is transmitted via contaminated food or water.

Shigella. All *Shigella* species and serotypes cause a dysentery that is characterized by severe, cramping abdominal pain accompanied by high temperatures, vomiting and diarrhea with blood and mucus. This disease is endemic under conditions of poor sanitation. There are at least 32 *Shigella* serotypes of which *Shigella sonnei* and the subgroups of *Shigella flexneri* account for more than 90% of all isolates from the human population. Shigellae are rarely found in animals other than humans. Most shigellosis epidemics are food-borne or spread by person-to-person contact. However, there have been a significant number of epidemics that were caused by poor-quality drinking water.

Vibrio. Of the many species present in water and soil, *V. cholerae* can produce a serious, acute intestinal disease that is characterized by sudden diarrhea with profuse, watery stools, vomiting, suppression of urine, rapid dehydration, fall of blood pressure, subnormal temperature, and complete collapse. Death may result within a few hours of onset unless prompt medical treatment is administered. *Vibrio cholerae,* the El Tor biotype, and Inaba and Ogawa serotypes are all pathogenic to humans. The spread of cholera may be through person-to-person contact, consumption of raw seafood contaminated by process water or food handling, and polluted drinking water.

Yersinia. *Yersinia enterocolitica* is another enteric pathogen that is a significant cause of acute gastroenteritis, involving the intestinal lymph nodes. The most common symptoms are fever and diarrhea with moderate dehydration. The fecal–oral route is the typical common mode of transmission. Like pathogenic *E. coli*, only certain strains of *Y. enterocolitica* that possess essential virulent factors are capable of causing human intestinal disease. Reservoirs of *Yersinia* are found in wildlife such as beavers, deer, and coyotes and can be transported in stormwater runoff through the watershed.

VIRAL AGENTS. Viral pathogens of concern in wastewater are those that infect humans, not farm animals or agricultural crops; however, dogs and farm animals may be carriers of human enteric viruses. These viruses will replicate in a single living susceptible cell of the infected host and eventually may be shed in the aquatic environment through mucus or fecal discharges. The infective dose is small, requiring only a few virus particles to make an individual sick. There is no evidence that they will multiply once they leave the human body but they do persist for some time in fecal cell debris traveling in polluted waters or in the wastewater environment. Viruses are beyond the resolution of the optical microscope but their structure can be studied under the electron microscope. They are separated into families on the basis of the type and form of the nucleic acid complex (DNA or RNA), size, shape, substructure, and mode of replication of the virus particle. Because enteric viruses are discharged in large numbers in the feces of infected hosts, they represent the most important virus group with regard to wastewater pollution and often reach their peak in summer and fall seasons.

Adenoviruses. These viruses have been associated with respiratory tract infections and conjunctivitis. These viruses, which have been isolated from wastewater and sludge, are more typically associated with epidemics involving a combination of fevers, colds, and sore eyes. The enteric adenoviruses, a subgroup of all adenoviruses, cause acute diarrhea disease and are associated with the fecal excretions of infants with gastroenteritis.

Astroviruses. These viruses have a unique star-shaped appearance when viewed by an electron microscope. They are found in wastewater but there are no known established reservoirs for these viruses. Several studies suggest that astroviruses can cause gastroenteritis, particularly among young children. Transmission of these organisms is by the fecal–oral route, like that of other enteric viruses.

Enteroviruses. The enteroviruses are divided into several groups that include polioviruses, coxsackieviruses, echoviruses and hepatitis A. These viruses typically infect humans, particularly infants and young children, causing a variety of illnesses ranging from aseptic meningitis to encephalitis, paralysis, pharyngitis, and pneumonia. They are highly transmissible, spread via the fecal–oral route. In temperate regions, enterovirus epidemics are more prevalent in the summer and fall. Enteroviruses are often found in both the influent and effluent from WWTPs.

COXSACKIEVIRUS. These viruses are transmitted through inhalation and ingestion of contaminated waters. Coxsackievirus A causes the common cold, aseptic meningitis, and conjunctivitis (eye infection). Coxsackievirus B causes several types of disease including heart disease.

ECHOVIRUS. These viruses cause a mild form of respiratory disease, but have occasionally been implicated in causing meningitis, diarrhea, fever, and rash. Echovirus and poliovirus are inactivated in raw wastewater more slowly than either *E. coli* or *Streptococcus faecalis.*

HEPATITIS A VIRUS. The hepatitis A virus is the causative agent of infectious hepatitis, a systemic disease primarily involving the liver. This pathogenic agent is typically associated with fecal–oral transmission through wastewater contamination and possesses the ability to withstand low levels of residual chlorine. Wastewater treatment personnel may have a potentially greater incidence of exposure to the hepatitis A virus because of their daily contact with wastewater.

POLIOVIRUS. This virus causes poliomyelitis, a disease that affects the central nervous system. Poliovirus can remain infectious for long periods of time in contaminated food and water. The primary mode of transmission is ingestion of feces-contaminated water containing the virus. Flies also serve as mechanical vectors of this disease (and for many other intestinal pathogens). The widespread use of polio vaccinations has greatly reduced the risk from exposure to this virus.

NORWALK VIRUS. The Norwalk virus is the cause of many gastroenteritis epidemics. This virus has been isolated from untreated or inadequately treated wastewater effluent entering lakes, rivers, and groundwater. Outbreaks of the illness have also been associated with faulty septic tank disposal systems, contaminated municipal water supplies, and recreational bathing waters of marginal quality. The Norwalk virus may be responsible for 23% of all waterborne outbreaks of gastroenteritis in the United States.

ROTAVIRUS. Rotaviruses are recognized as a common cause of acute viral gastroenteritis in all age groups. Outbreaks of this illness have been associated with wastewater-contaminated water resources. Raw wastewater and chlorinated wastewater effluent from activated-sludge plants treating domestic wastes have been shown to discharge high densities of these viruses each day.

PATHOGENIC PROTOZOANS. Pathogenic protozoans are shed in the feces of infected individuals, a variety of farm animals, and some wildlife. At this time, *Giardia lamblia* and *Crytosporidium* are the two pathogenic protozoans of greatest concern because of their passage through some unfiltered water supplies. Others like *Entamoeba histolytica* and *Cyclospora* are also threats that should be recognized. These organisms form a resistant cyst (much like a primitive egg) that enhances their ability to survive in a hostile wastewater environment and pass through treatment barriers (including disinfection processes) or settle in the sludge and remain viable. Their origins are in the intestinal tract of infected humans and a variety of wildlife. Infection frequently requires only a few viable cysts. See Chapter 11 for in-depth information on other parasites and their detection.

Cryptosporidium parvum. Individuals infected with this pathogen develop a profuse, watery diarrhea accompanied by abdominal pain, nausea, and anorexia. These symptoms are sometimes mistaken as being flulike. The disease is typically self-limiting, lasting for 1 to 3 weeks. In patients with acquired immune deficiency syndrome (AIDS) and others whose resistance to infection is impaired, it is much more serious and may lead to severe and lasting disability. The route of transmission is through ingestion of contaminated water with secondary spread by person-to-person contact in the family or nursery schools and other care institutions. The organism is widespread in the aquatic environment, derived primarily from farm animals (cattle, sheep, hogs, and goats), birds, other wildlife (deer, raccoons, foxes, coyotes, beaver, muskrat, rabbits, and squirrels), and pets (dogs and cats), with humans either only infected or serving as a reservoir of shedding oocysts. Infected humans may excrete as many as 1×10^{10} oocycts daily. The number of

Crytosporidium oocysts in raw wastewater can be as great as 1369 organisms/L (5180 organisms/gal) and this density was only reduced to 281 organisms/L (1063 organisms/gal) in treated wastewater (Madore, et al., 1987). *Cryptosporidium* has been found in association with rotavirus and *Giardia,* implying that outbreaks may at times involve multiple exposures to more than one pathogen in fecal contamination. Only since 1976 has *Crytosporidium* been linked to disease in humans, largely because of the difficulties in isolating and identifying the organism. This protozoan is very resistant to chlorination. Perhaps the most effective barrier to passage through water or wastewater treatment processes is to physically remove it with coagulation and filtration. Because *Cryptosporidium* oocysts remain viable in soil for months, it is likely that small numbers could be isolated from wastewater solids applied to soil.

Cyclospora. This protozoan is a recently recognized pathogen, first observed in 1979. It remains to be determined whether or not humans are the only natural reservoir for this agent. Ingestion of feces-contaminated water containing infectious oocysts of *Cyclospora* has been the only mode of transmission reported to date. There is some evidence to suggest that birds, insects, or dust particles could have played a part in several outbreaks involving rooftop storage of water supplies, cleaning a flooded basement, or eating strawberries, raspberries, and vegetables from fields possibly irrigated with poor-quality water. The disease produces a watery diarrhea with weight losses of up to 5 to 10%. Perhaps some of the unidentified watery diarrheas among adults is caused by the recently identified *Cyclospora* (AWWA, 1999).

Entamoeba histolytica. This protozoan causes amoebic dysentery and diarrhea, abscesses of the liver, and skin ulceration. This protozoan is a frequent intestinal parasite in humans, particularly in areas of the world where the disease is endemic. Dysentery epidemics of *En. histolytica* occur worldwide in developing countries and are largely waterborne; typically, they are a result of poor personal sanitation and deteriorating wastewater and water treatment infrastructures. Most occurrences reported in the past 25 years involve wastewater contamination of the distribution system for small, private water supplies. In the western world, improved sanitary sewerage systems have been largely responsible for the curtailment of waterborne amoebic dysentery, but caution needs to be exercised in solids handling and reuse of poorly treated wastewater effluent.

Giardia lamblia. This protozoan infects the small bowel (small and large intestine) and causes prolonged explosive, watery, foul-smelling diarrhea that may last for weeks with significant weight loss. Infected individuals may shed at least 1.1×10^6 cysts/g of stool. The carrier rate in different areas of the United States may range from 1.5 to 20%, depending on the community health and age group surveyed. *Giardia* concentrations from effluents of 11 WWTPs located across the United States ranged from 683 to 3750 cysts/L (Jakubowski, et al., 1991). Reservoirs of this pathogen include wild animals (beaver, muskrat, elk, and mule deer), farm animals (cattle and sheep), pets (dogs and cats) and humans. Important modes of transmission are drinking water from surface supplies receiving minimal treatment (*Giardia* is not inactivated by the conventional chlorine concentrations and contact times used) and through poor sanitation practices in daycare centers. It was not until the early 1950s that this protozoan was recognized as a pathogen and only in 1966 that wastewater-contaminated water was identified as an important vehicle of transmission.

*O*PPORTUNISTIC *PATHOGENS*

There is a wide variety of opportunistic pathogens that can present a potential risk to WWTP operators. These include the bacteria *Clostridium, Legionella,* and *Mycobacterium*; perhaps reovirus, and some of the fungi typically found in wastewater. Whenever the body's normal resistance is lowered as a result of colds, flu, or other illnesses; use of excessive amounts of analgesics that reduce normal levels of acidity in the stomach; or chemotherapy treatment, secondary infections from opportunistic pathogens may become a threat that must not be ignored.

OPPORTUNISTIC BACTERIA. The general population of bacteria in wastewater contains a variety of heterotrophic organisms, many of which have no significance to the general community health but may be a potential threat to some individuals under specific situations. Exposure to opportunistic pathogens in the outdoor environment may occur from aerosol inhalation or body contact with spraying wastewater. Under conditions of stress, weakened immune system, or through the general decreased physical health associated with aging, opportunistic organisms can quickly invade the human body to cause infections of the lung, skin, ear, eye, and urinary tract. When these situations arise, these bacteria can rapidly overcome the reduced natural body defense mechanisms and cause a variety of illness. See Chapter 9 for more information on indicator bacteria that may also include some opportunistic bacteria.

Clostridium. This genus consists of a large group of bacteria, many of which are ubiquitous in the water–soil environment. Unfortunately there are several species that should be of concern to wastewater operators. One of these is the organism that causes tetanus. When accidental cuts to the skin expose the body to soil, wastewater, or dirty equipment, entry of the ever-present *Clostridium tetani* to the blood stream may occur. Colonization by this organism results in the release of a toxin that attacks the central nervous system, ultimately leading to paralysis and muscle degeneration if not treated promptly. Immunization of wastewater personnel has helped to eliminate the occurrence of this disease among individuals exposed to the contaminated soil and water environment.

Clostridium difficile is another opportunistic *Clostridium* that is found in water and soil. This organism is a significant cause of a secondary illness (diarrhea) associated with changes in the natural flora of the intestinal tract resulting from antibiotic treatments and diarrhea among cancer patients who are treated with chemotherapeutic agents. Personnel who are receiving such treatments while working at the treatment plant during their therapy should be aware of the increased health risk.

Legionella. *Legionella pneumophila* is an important waterborne opportunistic pathogen that causes Legionnarie's disease in susceptible individuals exposed to contaminated aerosols from shower baths, air conditioner heat exchangers, fountains, and possibly wastewater treatment aeration processes. The respiratory disease results in a pneumonia with significant mortality rates among senior citizens. Pontiac fever, another illness caused by Legionellae, is a nonpneumonic, nonfatal, and self-limiting disease. Apparently, there is no human carrier state or reservoir for legionellae in warm-blooded animals.

Mycobacterium. Any of seven species of nontuberculous mycobacteria can be isolated from human fecal material. Waste from pig farms also contains mycobacteria and wastewater effluent may contain an approximate 10^4 organisms/100 mL. The pathological significance of these organisms is that inhalation of contaminated aerosols or contact through skin abrasions may result in colonization in the lungs and lymph nodes, skin lesions, or septicemia. Older people are more susceptible when in contact with aerosolized water containing these organisms during the summer months. Significantly, nontuberculous mycobacterial disease is the third most common opportunistic fatal infection in patients with AIDS.

OPPORTUNISTIC VIRUSES. Not too much is known about the occurrence of opportunistic viruses in the community. It is suspected that the ubiquitous abundance of such viral agents may put some individuals more at risk, but it has been difficult to document. Acquired immunities during early childhood probably protect many individuals later in life from these potential secondary invasions that cause complications during periods of illness.

Reovirus. These are perhaps the most often isolated viruses in the polluted aquatic environment, frequently found in raw wastewater and in various stages of the wastewater treatment train. They are known to be relatively resistant to the concentrations of disinfectants typically used in wastewater treatment. Reservoirs of reovirus are found not only in humans, but many other warm-blooded animals, including birds. Although they have been linked to a variety of mild respiratory and gastrointestinal infections common during childhood, there is no clear-cut association with specific human diseases. This virus may be an opportunistic agent that will attack an individual with a primary illness or a weakened immune system.

OPPORTUNISTIC FUNGI. These organisms are widely distributed in the environment and those isolated from wastewater are often associated with opportunistic diseases. Detection of a fungal agent in tissues or body fluids is essential because air contaminants in cultures or on lesions may result in misinterpretation. Most opportunistic fungi affect only the skin, hair, and nails, but some may infect any part of the body (systemic mycoses).

Aspergillus. Aspergilli are ubiquitous in the environment, often found growing on most organic materials, including grain, decaying vegetation, and soil. Composting is a significant source of potential infection for wastewater personnel. Of the approximately 200 species of *Aspergillus*, only four are typically encountered as a cause of disease: *Aspergillus fumigates, Aspergillus flavus, Aspergillus niger,* and *Aspergillus terreus.* It is important to note that aspergilli are the second most frequent cause of opportunistic fungal diseases. The spectrum of disease caused by these pathogenic strains is broad. Illnesses vary from allergic reactions to colonization of the respiratory tract or infections of the ear canal, nail, and sinuses. Respiratory infections may clinically resemble tuberculosis and other mycotic and bacteria diseases.

Candida. *Candida albicans* is an opportunistic pathogenic yeast that may occur in human feces at as great a concentration as 10^4 organisms/g. *Candida albicans* densities in raw wastewater range from a few thousand to as great as 25 000 organisms/L. Large numbers of this yeast are shed in active cases of diaper rash or by vaginitis. This organism has a high chlorine resistance compared with coliforms. At least seven species are involved in human infection but *Candida albicans* is by far the most frequently encountered yeast pathogen. Body contact and inhalation are considered routes for pathogen invasion. *Candida albicans* causes pulmonary infections, bronchitis, vaginitis, urethritis, and superficial infections of the skin and nails. *Candida* infections involving the mucous membranes of the mouth are referred to as thrush. Proper personal hygiene will reduce the risk of yeast infections to wastewater personnel.

TREATMENT CONTROL OF PATHOGENS

Most wastewater treatment processes remove and destroy pathogens to a varying degree. Moreover, the efficiency of a given process may vary considerably depending on the design of the plant, the nature of the influent and effluent, the season, and other related factors. Processes available for pathogen reduction from wastewater include absorption, coagulation, settling, and biological activity within the process reactor. Many of the bacteria common to wastewater contribute to the overall biological process by degrading the waste material and being antagonistic to pathogen survival.

PRIMARY TREATMENT. Primary treatment processes are designed to allow suspended solids to settle. Estimates of bacterial removal indicate 5 to 40% reductions from raw wastewater by this simple process. Sedimentation in an Imhoff tank for 2 hours brings about little or no virus reduction. However, primary treatment may remove large numbers of protozoa cysts and helminth eggs through sludge settling. It is unlikely that primary treatment alone can produce consistent reductions of great magnitude for most pathogenic organisms. Effective disinfection of the effluent to inactivate pathogens in primary effluent is often erratic because of the high disinfection demand. There are also concerns about odor problems generated and toxicity to fish plus their food chain organisms. Chlorination of nutrient-rich, primary effluent also leads to massive growth of coliform bacteria and extended persistence of *Salmonella* miles downstream of discharge.

TRICKLING FILTER PROCESS. Frequently, the stage of treatment beyond removal of solids involves a trickling filter that consists of a bed of material 0.9 to 3 m (3 to 10 ft) deep that supplies a large surface area through which the primary effluent passes. The trickling flow of primary settled wastewater through a bed of gravel, stone, cinders, or slag brings about intimate contact between the primary waste effluent and the organisms that naturally thrive on the nutrients in the effluent. A large variety of organisms including bacteria, fungi, and yeast develop the filter slimes that participate in the stabilization process and, through competition, reduce the presence of pathogens. Trickling filters reduce *Salmonella paratyphi B* densities by 84 to 99%, *Mycobacterium tuberculosis* populations by 66%, enteric viruses by 24 to 60%, tapeworm ova by 18 to 70%, and cysts of *Endameba histolytica* by 88 to 99% (Geldreich, 1978).

AEROBIC SLUDGE PROCESS. Most aerobic and anaerobic sludge treatment processes significantly reduce the number of pathogens. Aerobically digested sludge has been shown to reduce both enteric bacteria and viruses. *Salmonella* spp. and *Shigella* spp. have been reported to be 90 to 99% lower in effluent than in incoming raw wastewater. Enteroviruses can be reduced by 90% through the activated-sludge process as a result of absorption of the virus particles onto settling sludge floc.

   Although the process removes large numbers of bacteria and viruses, there may be extended survival of residual pathogens in the sludge, so care must be exercised in the storage and disposal of this byproduct of treatment. Solids dewatering and high temperature drying are important control measures.

ANAEROBIC DIGESTION PROCESS. Anaerobic digestion is widely used to stabilize concentrated organic solids removed from primary settling tanks, biological filters, and the activated-sludge process. A multiyear study of four WWTPs isolated a total of 297 enteroviruses from 307 samples from aerobic and anaerobic digestion processes (Hamparian, et al., 1985). Unfortunately, anaerobic digestion does not inactivate significant numbers of ova from various intestinal parasites. Although the concentrations of such ova may be slowly reduced during practical retention periods, significant numbers remain after treatment. Viable eggs have been recovered following digestion for as long as 3 to 6 months. However, this wastewater treatment process is effective in reducing cysts of *E. histolytic*. Density reductions of *M. tuberculosis* by anaerobic digestion may be on the order of 69 to 90%. Removals of *S. typhi* range from 24 to 92.4%, depending on the length of retention (Geldreich, 1978).

WASTE STABILIZATION LAGOONS. Large, shallow lagoons are used in some localities as an economical wastewater treatment method. A waste stabilization lagoon is an artificially created pond or lagoon designed to treat waste by biological, chemical, and physical processes typically referred to as natural self-purification. Adequate retention time is essential to optimize waste degradation and microbial inactivation.

These wastewater treatment lagoons reduce salmonellae by 70 to 99.9% when the retention time is adequate. Reduction of virus numbers by 68 and 92% was reported in two studies on lagoons that had 20- and 30-day retention periods. *Schistosoma* (the blood fluke flatworms), and other parasites that go through a motile aquatic stage, persist for some time under favorable conditions (Berg and Metcalf, 1978; Geldreich, 1972 and 1978; and Kott, et al., 1978).

WASTEWATER DISINFECTION. Disinfection is an important process in suppressing the transmission of pathogens through wastewater treatment. Most enteric bacteria and viruses are destroyed by chlorine, ozone, and UV light disinfection , provided contact time is adequate and there is low turbidity and excellent nutrient removal in the process waste. Widespread use of wastewater chlorination has played an important role in preventing more frequent disease outbreaks in the aquatic environment. However, poor treatment of waste before careful application of chlorine has often led to ineffective microbial reductions, foul odors in receiving waters, and bacterial regrowth downstream. Even with proper disinfection of a high-quality effluent, some pathogens will survive. For example, Mycobacterium species have been shown to be more resistant to chlorine than various common heterotrophic bacteria. The Norwalk virus and hepatitis A have also been reported to be more resistant to chlorine than other enteroviruses. Other factors that influence the efficiency of disinfection include pH, temperature, disinfectant dosage, particulates, and fecal cell clumping.

INDIVIDUAL ONSITE DISPOSAL SYSTEMS. A septic tank is simply a receptacle buried in the ground to treat the wastewater from an individual home or a small business establishment. Wastewater from the home flows into the septic tank where bacteria degrade the organic matter. The resulting effluent is periodically displaced to a subsurface gravel- or sand-filled seepage pit or drainage field for diffusion into the surrounding soil. With careful operation, these small individual systems can be effective in reducing the microbial content of raw wastewater. Unfortunately, these treatment systems are often poorly constructed, receive little maintenance, and are overloaded by the householder so that the wastewater receives little natural purification treatment and limited microbial reductions occur. Reduction of bacterial densities in the effluent often range from 25 to 75%. Many septic tanks contain enteroviruses. In one instance, a septic tank still contained viable polioviruses 6 months after a child from that home contracted poliomyelitis (Geldreich, 1978).

REFERENCES

American Water Works Association (1999) *Manual of Water Supply Practices: Waterborne Pathogens.* AWWA Manual M48, Denver, Colo.

Berg, G., and Metcalf, T. G. (1978) Indicators of Viruses in Water. In *Indicators of Viruses in Water and Food.* Ann Arbor Sci., Ann Arbor, Mich.

Cullaghan, P., and Brodie, J. (1978) Laboratory Investigation of Sewer Swabs Following the Aberdeen Typhoid Outbreak of 1964. *J. Hyg.* (G.B.), **66**, 489.

Geldreich, E.E. (1972) Water-Borne Pathogens. In *Water Pollution Microbiology.* R. Mitchell (Ed.), Wiley & Sons, New York, 207.

Geldreich, E.E. (1978) Bacterial Populations and Indicator Concepts in Feces, Sewage, Stormwater and Solid Wastes. In *Indicators of Viruses in Water and Food.* G. Berg (Ed.), Ann Arbor Science Pub., Ann Arbor, Mich., 51.

Geldreich, E.E. (1996) *The Worldwide Threat of Waterborne Pathogens in Water Quality in Latin America: Balancing the Microbial and Chemical Risks in Drinking Water Disinfection.* ILI Press, Washington, D.C., 19.

Geldreich, E.E., and Van Donsel, D.J. (1970) Salmonellae in Fresh Water. *Proc. Natl. Specialty Conf. Disinfection,* Am. Soc. Civ. Eng., 495.

Hamparian, V.V.; Ottonlenghi, A.C.; and Hughes, J.H. (1985) Enteroviruses in Sludge, Multiyear Experience with Four Wastewater Treatment Plants. Appl. Environ. Microbiol., **50**, 28.

Harvey, R.W.S.; Price, T.H.; Foster, D.W.; and Griffith, W.C. (1969) Salmonella in Sewage. A Study of Latent Human Infections. *J. Hyg.* (G.B.), **67**, 517.

Jakubowski, W; Sykora, J.L.; and Sorber, C.A. (1991) Determining Giardiasis Prevalence by Examination of Sewage. *Water. Sci. Technol.,* **24**, 173.

Kott, Y.; Ben-Ari, H.; and Betzer, N. (1978) Lagooned, Secondary Effluents as Water Source for Extended Agricultural Purposes. *Water Res.* (G.B.), **12**, 1101.

Madore, M.S., et al. (1987) Occurrence of *Crytosporidium* Oocysts in Sewage Effluents and Selected Surface Water. *J. Parasitol.,* **73**, 702.

Swerdlow, D.L.; Bradley, M.D.; Woodruff, A.; et al. (1992) A Waterborne Outbreak in Missouri of *Escherichia coli O157:H7* Associated with Bloody Diarrhea and Death. *Ann. Int. Med.,* **117**, 10, 812.

SUGGESTED READINGS

American Public Health Association (1995) *Control of Communicable Diseases Manual.* 16th Ed., A.S. Benenson (Ed.), Washington, D.C.

American Public Health Association; American Water Works Associations; and Water Environment Federation (1995) *Standard Methods for the Examination of Water and Wastewater.* 19th Ed., Washington, D.C.

American Society for Microbiology (1985) *Manual of Clinical Microbiology.* 4th Ed., E.H. Lennette; A. Balows; W.J. Hausler, Jr.; and H.J. Shadomy (Eds.), Washington, D.C.

Berg, G.; Bodily, H.L.; Lennette, E.H.; Melnick, J.L.; and Metcalf, T.G. (Eds.) (1976) *Viruses in Water.* American Public Health Association, Washington, D.C.

Bitton, G. (1980) *Introduction to Environmental Virology.* Wiley & Sons, New York.

Blacklow, N.R., and Cukor, G. (1981) Viral Gastroenteritis. *New Engl. J. Med.,* **304**, 397.

Blacklow, N.R., and Cukor, G. (1982) Norwalk Virus: A Major Cause of Epidemic Gastroenteritis. *Am. J. Publ. Health,* **72**, 1321.

Erlandsen, S.L., and Meyer, E.A. (Eds.) (1984) *Giardia and Giardiasis: Biology, Pathogenesis, and Epidemiology.* Plenum Publishing, New York.

Geldreich, E.E. (1996) Pathogenic Agents in Freshwater Resources. *Hydrolog. Process.,* **10**, 315.

Hejkal, T.W; Smith, E.M.; and Gerba, C.P. (1984) Seasonal Occurrence of Rotavirus in Sewage. *Appl. Environ. Microbiol.,* **47**, 558.

Janoff, E.N., and Barth Reller, L. (1987) *Cryptosporidium* Species, A Protean Protozoan: Minireview. *J. Clin. Microbiol.,* **25**, 967.

Lin, S.D. (1985) *Giardia lamblia* and Water Supply. *J. Am. Water Works Assoc.,* **77**, 40.

O'Leary, W. (Ed.) (1989) *Practical Handbook of Microbiology.* CRC Press, Inc., Boca Raton, Fla.

Rose, J.B.; Cifrino, A.; Madore, M.S.; et al. (1986) Detection of *Cryptosporidium* from Wastewater and Freshwater Environments. *Water Sci. Technol.,* **18**, 233.

Ward, R. (1994) *Cyclospora*: A Newly Identified Intestinal Pathogens of Humans. *Clin. Infect. Dis.,* **18**, 620.

Chapter 11
Wastewater Parasites

CHARACTERIZATION

The uncontrolled transmission of many diseases has been effectively reduced in the United States by the use of wastewater conveyance and treatment systems. Unfortunately, these systems can expose wastewater personnel to relatively large numbers of disease-causing agents. Although the incidence for acquiring disabling parasitic infections is less than 1% for wastewater personnel (Gerardi and Zimmerman, 1997), this exposure presents a potential health risk.

Numerous parasitic organisms are found in domestic wastewater (Table 11.1). Because the infective eggs of flatworms (flukes and tapeworms), roundworms, and the infective cysts of protozoa are more dense than water, many are removed from the waste stream by primary clarification. Primary sludge, therefore, can contain a relatively high concentration of a variety of disease agents. The eggs of the roundworm, *Ascaris*, for example, may be present at a concentration as high as 198 000 eggs per kilogram of primary sludge. Eggs and cysts that pass through primarily clarifiers can also be removed from the waste stream in secondary treatment processes by entrapment or association with biological solids, such as activated-sludge flocs and trickling filter biofilms.

Table 11.1   Parasites found in sludge samples.

Parasite found	Probable identity	Definitive host
Nematoda		
Ascaris eggs	*Ascaris lumbricoides*[a]	Humans
	Ascaris suum[a]	Pigs
Ascaridia-like eggs	*Ascaridia galli*	Domestic poultry
	Heterakis gallinae	Domestic poultry
Trichuris trichiura	*Trichuris trichiura*	Humans
	Trichuris suis[b]	Pigs
Trichuris vulpis eggs	*Trichuris vulpis*	Dogs
Toxascaris-like eggs	*Toxascaris leonina*	Dogs and cats
Parascaris equorum eggs	*Parascaris equorum*	Horses
Gongylonema-like eggs	*Gongylonema neoplasticum*	Rats
	Gongylonema pulchrum	Cattle, pigs, etc.
Toxocara eggs	*Toxocara canis*[c]	Dogs
	Toxocara cati	Cats
Trichosomoides-like eggs	*Trichosomoides crassicauda*	Rats
	Anatrichosoma buccalis	Oppossums
Cruzia-like eggs	*Cruzia americana*	Oppossums
Capillaria spp. eggs	*Capillaria hepatica*	Rats
(three or more types)	*Capillaria gastrica*	Rats
	Capillaria spp.	Domestic poultry
	Capillaria spp.	Wild birds
	Capillaria spp.	Wild mammals (oppossums, raccoons, and so on)
Platyhelminthes		
Hymenolepis diminuta eggs	*Hymenolepis deminuta*	Rats
Hymenolepis nana eggs	*Hymenolepis nana*[d]	Humans and rodents
Hymenolepis sp. eggs	*Hymenolepis* spp. (possibly more than one species)	Domestic and wild birds
Taenia sp. eggs	*Taenia saginata*[e]	Humans
	Taenia pisiformis[e]	Cats
	Hydratigera taeniaeformis[e]	Dogs
Diphyllobothrium-like eggs	*Diphyllobothrium latum*	Humans, dogs, bears
	Diphyllobothrium spp.	Dogs, bears, birds
Schistosoma mansoni	*Schistosoma mansoni*	Humans
Spirometra-like eggs	*Spirometra mansonoides*	Dogs and cats
Acanthocephala		
Acanthocephalan eggs	*Macracanthorhynchus hirudinaceus*	Pigs
Protozoa		
Giardia cysts	*Giardia lamblia*	Humans
	Giardia spp.	Dogs, cats, mammals
Coccidia oocysts	*Isospora* spp.	Dogs, cats
	Eimeria spp.	Domestic and wild birds, mammals

[a] Eggs of *A. lumbricoides* and *A. suum* are indistinguishable.
[b] *T. suis* eggs were probably only rarely seen.
[d] *Toxocara* eggs were probably mostly *T. canis*.
[d] Recent scientific name change of *Hymenolepis nana* is *Vampirolepis nana*.
[e] Eggs of these worms were indistinguishable.

DEFINITIONS. A parasite lives on (ectoparasite) or in (endoparasite) another organism of a different species from which it derives its nourishment. The organism is called the parasite's host. The parasite, however, typically does not kill its host because the life of the parasite would also be terminated. The host in which the parasite reaches sexual maturity and reproduction is termed the definitive host. If no sexual reproduction occurs in the life of the parasite, such as in protozoans like *Giardia*, the host believed to be the most important to it is arbitrarily called the definitive host (often a human or other mammal). An intermediate host is one in which some developmental or larval stage of the parasite occurs, but in which it does not reach maturity. When a parasite enters the body of a host and does not undergo any development but continues to stay alive and be infective to a definitive host, the host is called a paratenic or transport host. Paratenic hosts are often useful or necessary for completion of the life cycle of the parasite because they may bridge the gap between the intermediate and definitive hosts (Schmidt and Roberts, 1989).

LIFE CYCLE. Most parasites are obligate parasites; that is, they must spend at least part of their lives as parasites to survive and complete their life cycles. However, many obligate parasites have free living stages outside of a host, including some periods of time as a protective egg (ova) or cyst. These resistant ovum or cyst stages may show up in wastewater or sludge samples. Facultative parasites are not typically parasitic but can become so, at least for a time, when they are accidentally eaten or enter a wound or other body opening. When a parasite enters or attaches to the body of a host species different from its normal one, it is called an accidental or incidental parasite (Schmidt and Roberts, 1989, and Zaman and Keong 1982). The life cycle of a typical parasite is shown in Figure 11.1.

IDENTIFICATION. Parasitology texts (Schmidt and Roberts, 1989, and Zaman and Keong 1982) and laboratory manuals (Cable, 1977, and Meyer et al., 1988) are excellent sources for descriptive characteristics of parasitic helminth ova and protozoan cysts. Some typical forms of the helminth ova are shown in Figure 11.2 (Meyer et al., 1988). Color atlases by Fox et al. (1981) and Berk and Gunderson (1993) contain chapters that deal specifically with wastewater parasites. The pictorial atlas of Ash and Orihel (1984) shows many human ova and cysts, and the text by Olson (1974) deals specifically with parasites of wild and domestic animals (which may contribute ova or cysts to a municipal wastewater treatment plant).

OCCURRENCE IN WASTEWATER AND SLUDGE

The number and variety of parasitic forms present in wastewater or sludge depend heavily on the origin of the influent. Broad taxonomic groups (phyla)

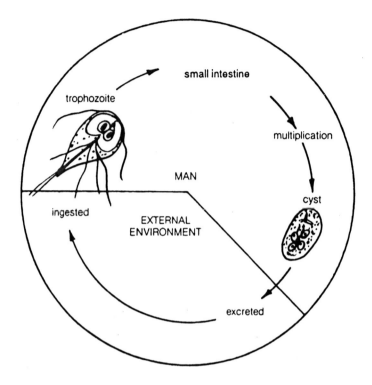

Figure 11.1   Life cycle of *Giardia lamblia*.

with ova or cysts that may be present in wastewater and sludge include the protozoa (such as *Entamoeba; Giardia; Cryptosporidia;* and the helminths, including nematodes [roundworms such as *Ascaris* and *Trichuris*], platy-helminthes [flatworms such as the trematode fluke *Fasciola* and cestode tapeworms such as *Taenia* and *Hymenolepis*], and the acanthocephala [spiny-headed worms such as *Macracanthorhynchus*]).

   Comprehensive data reviewing the occurrence of animal and human parasites in the wastewater and sludge of U.S. treatment plants is limited. However, a survey (Reimers et al., 1981 and 1986) of wastewater and sludge samples from all phases of treatment from 75 municipal plants (located in Alabama, Florida, Louisiana, Minnesota, Mississippi, New York, Ohio, Texas, and Washington) to assess the presence and densities of parasites revealed

- Resistant stages of 23 types of parasites found in samples, which may represent more than 30 different human and animal parasites (see Table 11.1). Although the densities may vary, parasites are distributed throughout all of the continental United States.
- The four most often found parasites were the nematodes, *Ascaris* spp., *Trichuris*, and *Toxocara* spp.
- The total number of parasite ova recovered varied according to the source of sludge and season of the year and averaged 14 000 eggs/kg (dry weight) of sludge.

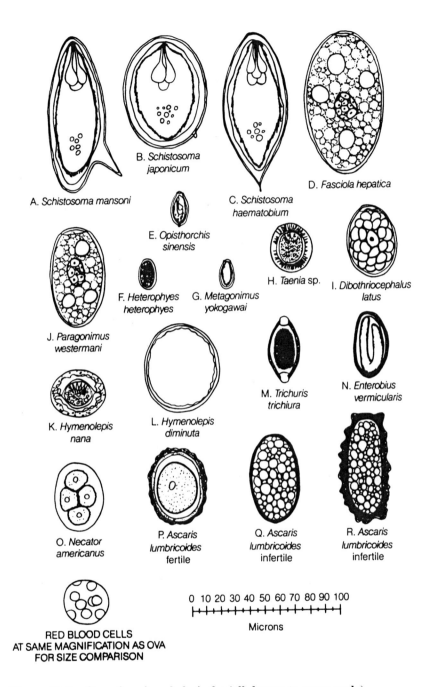

A. *Schistosoma mansoni*
B. *Schistosoma japonicum*
C. *Schistosoma haematobium*
D. *Fasciola hepatica*
E. *Opisthorchis sinensis*
F. *Heterophyes heterophyes*
G. *Metagonimus yokogawai*
H. *Taenia sp.*
I. *Dibothriocephalus latus*
J. *Paragonimus westermani*
K. *Hymenolepis nana*
L. *Hymenolepis diminuta*
M. *Trichuris trichiura*
N. *Enterobius vermicularis*
O. *Necator americanus*
P. *Ascaris lumbricoides* fertile
Q. *Ascaris lumbricoides* infertile
R. *Ascaris lumbricoides* infertile

RED BLOOD CELLS
AT SAME MAGNIFICATION AS OVA
FOR SIZE COMPARISON

0 10 20 30 40 50 60 70 80 90 100
Microns

Figure 11.2   Ova of various helminths (all drawn to same scale).

- The percentage of the total parasite ova in the sludge samples that were viable (capable of causing an infestation) ranged from 0 to 100%, but was typically greater than 45% for primary sludge and 69% for treated biosolids.

- Primary and secondary undigested sludge samples were found to contain, in order of decreasing average densities: 9700 *Ascaris* spp. ova, 1200 *Toxocara* spp. ova, 800 *T. trichiura* ova, and 600 *T. vulpis* ova/dry kg of sample.
- The average numbers of these parasites in stabilized biosolids samples were 9600 *Ascaris* spp. ova, 700 *Toxocara* spp. ova, 2600 *T. trichiura* ova, and 700 *T. vulpis* ova/dry kg of sample.

Other typical parasites that have been identified in wastewater and sludge include (Fox et al., 1981)

- Nematodes: *Enterobius vermicularis, Strongyloides steicorallis, Trichinella spiralis,* and *Necator americanus*;
- Trematodes: *Clonorchis sinensis, Schistosoma mansoni, Fasciola hepatica,* and *Heterophyes literophyes*;
- Cestodes: *Taenia solium*; and
- Protozoa: *Entamoeba histolytica, Balantidium coli, Nagleria fowleri,* and *Cryptosporidia* spp.

INACTIVATION AND DESTRUCTION

Peculiarities in life cycle and presence of protective structures–shells surrounding worm eggs or protozoan cysts allow some parasites to be more resistant to treatment than others. Also, the survival rates of the parasitic stages are affected by the wastewater or solids treatment processes to which they are subjected. Studies of the effectiveness of treatment have shown that (Reimers et al., 1981 and 1986)

- During digestion, destruction of resistant parasite ova is primarily caused by temperature and not to a specific digestion process. Aerobic or anaerobic digestion inactivated parasite ova at temperatures greater than 55 °C (131 °F) in 2 hours and at 45 °C (113 °F) within days but only retarded ova development at temperatures lower than 45 °C. *Ascaris* and *Toxocara* ova were the most resistant.
- Lime treatment of sludges did not produce consistent inactivation of *Ascaris*.
- The results of ammonification treatment were inconclusive, especially with *Ascaris*.
- Ultrasonication was effective in destroying *Toxocara* ova at 49 kHz during a 6-minute exposure, but was not effective in destroying *Ascaris* eggs.
- During a 15-day period of mesophilic anaerobic digestion, 23% of *Ascaris* ova were inactivated, whereas *Trichuris* ova were not affected.

Wastewater Biology: The Microlife

- Aerobic digestion inactivated 38% of the *Ascaris* eggs and 11% of *Trichuris* eggs.
- *Ascaris* ova aerobically digested along with sludge were noninfective by 16 months when stored at 25 °C (77 °F) or in the ground at 4 °C (39 °F).
- *Ascaris* ova anaerobically digested with sludge were noninfective by 16 months when stored at 25 °C, but were still infective after 22 months when stored at 4 °C.
- *Trichuris* spp. ova in digested sludge seemed to be inactivated by storage at 25 °C in the ground after 16 months, but could be viable even after 25 months when stored at 4 °C.

Typically, each process that exposes a parasite to a different or hostile environment may shorten its survival time. For example, aerobic digestion, followed by anoxic and anaerobic treatment processes exposes a parasite to several hostile environments. If composting then follows (especially when temperatures reach 60 °C (140 °F), conditions for the destruction and reduced viability of parasites is further intensified (Pedersen, 1981).

MICROSCOPIC EXAMINATION

SAMPLE COLLECTION AND CONCENTRATION. To obtain an accurate profile of the parasites present in wastewater and sludge, samples must be obtained and examined from all sections of the treatment process. The influent provides a survey of what is coming into the plant and the effluent provides a survey of what leaves the plant. However, the examination of sludge from various parts in the plant provides the most meaningful listing and enumeration of parasites. Most ova or cysts are removed in treatment processes through sludge settling. It should be noted that the presence of ova or cysts in settled sludge does not necessarily mean that they are viable.

Sludge with solids concentrations of 2 to 3% are most useful for processing and examining parasitic ova and cysts. Therefore, depending on the source, the sample will have to be concentrated or diluted. In either case, sample volumes should be recorded so that counts can be related to the initial sample if enumeration is done.

If dilution is necessary, distilled or filtered water should be used to prevent adding organisms to the sample. A final sample volume of 500 mL of concentrated or diluted wastewater or sludge is desirable for processing. Typically, 40 L of influent is often necessary to obtain 500 mL of a 2 to 3% solids concentrated sample.

Concentration of the influent or effluent can be done by either continuous-flow centrifugation (1000 r/min for 1 to 2 minutes) or by simple sedimentation in straight-sided containers (such as Imhoff cones). The recommended time for sedimentation is 24 hours, but it can range from 12 to 24 hours (Fox et al., 1981).

The supernatant fluid in the centrifugation tube or sedimentation container should be aspirated or siphoned off. Decanting might result in loss of solids and should be avoided. If present, ova or cysts will be in the solids portion of the sample and will require further processing. Mixed liquor or digested sludges with a solids concentration of 2 to 3% can be processed directly. Because centrifugation may damage delicate cyst stages, concentration by sedimentation is preferred.

Immediate further processing of the concentrated sample for identification and enumeration of parasitic cysts or eggs is recommended. However, samples may be stored in a refrigerator for a period not to exceed 24 hours. If a longer delay before examination is necessary, then the whole sample (or a portion, if the sample is large) must be preserved with, either neutral 5% formalin, merthiolate–iodine–formalin (MIF), or polyvinyl alcohol (PVA) fixative (see Appendix A for procedure). If possible, fresh, unfixed samples should be examined because some cysts or eggs can be damaged by formalin treatment.

SEPARATION OF PARASITIC STAGES FROM DEBRIS. Two basic methods are used for isolating parasitic stages from wastewater or sludge: flotation and sedimentation analysis. With flotation analysis, many helminth ova and protozoan cysts will be separated from debris because they will float in high-density solutions (that is, with a specific gravity >1.18). However, some ova of tapeworms and nematodes are too dense to float. In this case, a sedimentation analysis is done. A comprehensive wastewater and sludge survey will involve both techniques.

Some of the typical compounds used as media in flotation techniques are sucrose, zinc sulfate, sodium or potassium dichromate, sodium chloride, sodium salicylate, and cesium chloride. When choosing a flotation media, the nature of the parasitic stage to be recovered and the type of the wastewater or sludge to be examined must be considered. The two most widely used flotation techniques are sucrose flotation and zinc sulfate flotation. The most common sedimentation technique is formalin–ether sedimentation (Cable, 1977; Fox et al., 1981; Meyer et al., 1988; and Pritchard and Kruse, 1982).

Zinc Sulfate Flotation. The following technique can be used to analyze wastewaters and sludges:

1. If concentrated samples contain large debris, pass a known amount of sample through a series of sieves (with 0.84, 0.42, and 0.30 mm openings) or through a stemless, ribbed glass funnel containing cheesecloth. Rinse the funnel or sieves with distilled or filtered water.
2. Pour filtered samples into 50-mL, round-bottom centrifuge tubes. The number of tubes needed depends on the sample volume. Fill each tube three-fourths full. Centrifuge the tubes at 1500 r/min for 3 minutes.
3. Aspirate the supernatant. Fill the tube approximately one-half full with filtered water and resuspend the sediment (use a glass rod, put a

cork on the tube, and shake vigorously or use a vortex mixer). Centrifuge the tubes again at 1500 r/min for 3 minutes.

4. Repeat Step 3 until the supernatant is fairly clear. Three times should be sufficient.

5. Transfer all pellets or solids at the bottom of the tube to 15-mL conical centrifuge tubes. Do not add more than 4 mL of pellet to any one tube.

6. In each 15-mL conical centrifuge tube, mix by vortexing each 5-mL pellet portion with 7 to 10 mL of zinc sulfate solution (a specific gravity of 1.18 if the sample is not preserved in formalin and 1.20 if the sample is preserved in formalin [see Appendix A]).

7. Add one to three drops of Lugol's iodine (see Appendix A), or other appropriate stain (see subsequent section), and bring the volume to approximately 15 mL with the zinc sulfate solution. Note that there is a preference not to stain until it is needed to make slides (after step 9, below).

8. Centrifuge (in a swinging bucket rotor) at 1500 r/min for 3 minutes (800 r/min for 5 minutes may be preferable) without braking (that is, let the centrifuge stop without using the mechanical brake). Place the tubes in an upright tube rack.

9. Transfer the sample to a slide using one of the following methods:
   • Slowly (drop by drop) fill the tube to the brim with zinc sulfate without running over and allow tube to remain static for 2 to 3 minutes. Next, touch the meniscus with a No. 1, 22-mm × 22-mm cover glass and place drop-side down on a clean glass slide and seal with melted vaspar (1 part petroleum jelly to 1 part paraffin, or use clear nail polish).
   • Using a wire loop, place several loopfuls of the surface film on a clean microscope slide. Add cover glass and seal as above. If a counting chamber is to be used, then a wire loop or micropipette is typically used for transfer of sample to slide.

10. To perform the microscopic examination, scan the whole slide (first at 100X then at 400X or 450X), identify or count the number of ova or cysts. Verification of some cysts may require 1000X. A phase-contrast microscope is desirable.

Sucrose Flotation. Use of this technique requires preparation of continuous sucrose gradients on commercial gradient formers, or formation of discontinuous sucrose gradients if a gradient former is not available. The gradient former consists of 12 mL of 80% Sheather's sucrose solution (specific gravity = 1.40, see Appendix A) and 8 mL of distilled water prepared in 50-mL, round-bottom centrifuge tubes. The specific gravity in the tube ranges from 1.3 at the bottom to 1.0 at the top. Discontinuous sucrose gradients are formed from 80, 50, 34, and 20% dilutions of stock Sheather's sucrose solution. Starting with the most dense solution (80%) and finishing with the least dense (20%), 5-mL portions of these solutions are carefully and slowly layered on top of one another in a 50-mL, round-bottom centrifuge tube.

Well-defined interfaces must be formed and, when completed, a glass rod is pushed slowly through the interfaces and then removed to break the interfaces. (Note that the gradient should not be stirred.) The parasitic stages can be separated as follows:

1. If a concentrated or diluted sample contains a lot of debris, treat as in step 1 of zinc sulfate technique.
2. Place 10- or 20-mL aliquots on separate, fresh 20-mL continuous sucrose gradients (see above).
3. Centrifuge at 2900 r/min for 10 minutes.
4. Aspirate the upper 15 mL of sucrose gradient using a syringe and transfer to 15-mL conical centrifuge tubes.
5. Centrifuge at 500 r/min for 5 minutes without mechanical breaking.
6. Prepare and view microscope slides as described in steps 9 and 10 of the zinc sulfate technique.

Formalin–Ether Sedimentation. The basis of formalin–ether sedimentation is chemical separation of organic debris to the top and settling of parasitic stages to the bottom of a centrifuge tube (Steer et al., 1974). This technique is reported to yield good recovery rates (95 to 100%). The recommended procedure is as follows (note that it is extremely important to work in well-ventilated areas when using formalin or ether):

1. Place 20 g of sludge directly, without prior dilution, in 200 mL of warm distilled water containing 0.5 g of an anionic detergent and mix in a blender for 45 seconds.
2. Pour 30 mL of the homogenate through a sieve with 0.84-mm openings and rinse with 20 mL of distilled water.
3. Mix the filtered material and pour 12.5 mL into a 50-mL, round-bottom glass centrifuge tube (plastic tubes cannot be used with ether).
4. Top the tubes with distilled water, stir, and spin in a centrifuge at 2000 r/min for 2 minutes.
5. Aspirate the supernatant and repeat washing (step 4) with distilled water until supernatant remains clear (two to three washings).
6. Remove the final supernatant and add 7 mL of 10% neutral formalin to each tube and mix thoroughly (shake or use a vortex mixer).
7. Add 3 mL of cold ethyl ether and shake vigorously for 30 seconds.
8. Centrifuge at 2000 r/min for 2 minutes.
9. Examine the centrifuge tube and observe a layer of ethyl ether on top followed by a layer of debris or fat, followed by a layer of formalin sitting above the sediment, which may contain the ova and cysts. Loosen the plug of organic debris with an application stick and discard this and all fluid above the sediment.
10. Mix the sediment and examine microscopically drop by drop as described in step 10 of the zinc sulfate technique.

STAINING METHODS. Lugol's iodine stain is universally used for staining protozoa cysts and helminth ova (see Appendix A). Helminth eggs stain well because the iodine is able to penetrate through thick shells. Iodine stain also gives good contrast to nuclear detail of protozoan cysts. Other typically used stains include buffalo black, methylene blue, or MIF. Most recently, immunofluorescence staining has become available for a variety of protozoan parasites.

ENUMERATION METHODS. General surveys for the presence or absence of parasitic stages do not require that quantitative counts be made. Simple scanning of the microscope slides and identifying ova or cysts are all that is required. Semiquantitative estimates of density can be made by scanning and counting parasites in the whole slide area from all slides prepared from an initial volume of sample. Computing the number of ova or cysts per volume requires calculations from the initial volume and concentration and dilution volumes. A number of studies report the number of cysts or ova per gram of dried sludge. This is done by splitting a known volume of the initial sample (100 mL) and oven-drying this volume at 80 to 105 °C (176 to 221 °F). The number of ova per gram is then calculated by multiplying the weight of a milliliter of sample (that is, mL/g) by the number of ova per milliliter.

The most quantitative counts of ova or cysts require the use of a counting chamber. A number of commercially available counting devices designed for other uses can also be used to enumerate ova or cysts. These include the Sedgwick-Rafter counting cell, Palmer-Maloney counting cell, McMaster counting cell, and the hemacytometer. Application of the Sedgwick-Rafter and Palmer-Maloney counting cells, which were specifically designed for plankton counting, are outlined in *Standard Methods* (APHA et al., 1995). These cells are specifically calibrated to hold 1.0 and 0.1 mL of sample, respectively.

The McMaster counting chamber consists of two microslides separated by 1.5-mm spaces. The top slide has two grids that are 1.0 cm^2. Each square is divided by parallel lines into six equal divisions to guide the user while scanning the area of the grid. The number of ova counted in both chambers is then divided by a factor of two and multiplied by 6.67 to account for the chamber depth. This yields the number of organisms per milliliter of sample. A disadvantage of the McMaster chamber is that it cannot be used easily for counts where magnifications greater than 100X are required for identification (Fox et al., 1981). This limitation of the technique will not allow identification of smaller cysts or ova. The hemacytometer, which is routinely used to make blood cell counts, overcomes this limitation (Gerardi et al., 1983). Use of the chamber is described in Appendix A.

*R*EDUCTION OF RISK

Although wastewater personnel become infected with disease agents that are often found in wastewater and solids, determining whether an infection was a

---

result of occupational exposure often is difficult. For example, it is difficult to determine if giardiasis experienced by a treatment plant worker is the result of an infestation of *Giardia* cysts from contamination from primary sludge or because the worker ate a salad prepared by a person whose hands also changed an infant's contaminated diaper. Regardless of the occupational risk assigned or perceived, risk can be greatly reduced with the use of proper Lydiene measures:

(a) Hand washing. Transmission of disease agents can be reduced significantly by never eating, drinking, smoking, or using tobacco products or gum before proper hand washing. The use of designated areas for these activities can also reduce the risk of transmission. Hands should be washed first with a detergent or soap that will weaken or remove the potential disease agent and then with an antimicrobial product. A nail brush should be used to remove solids from under fingernails.

(b) Gloves. To reduce the risk of infection, gloves should be worn while working and should be disposed of, or thoroughly washed after work. Disposable latex gloves may be used when sampling or performing jobs that produce little risk of abrasion. Heavy reinforced gloves should be worn on jobs when hands are susceptible to abrasion. The top of the glove should never be submerged and where it is not practical to wear gloves, hands should be washed immediately when a job is finished.

(c) Clothing. Direct contact with disease-causing agents or soiled clothing can be reduced by using protective clothing and equipment, including the use of goggles for eye protection.

(d) Respirators. The inhalation of disease agents in work areas with relatively large amounts of aerosols and dust may be prevented by using a properly fitted dust respirator.

(e) Sample bottles. To reduce contamination of hands and clothing from spills and broken sample bottles, wastewater and sludge samples should be collected in plastic, wide-mouth bottles with tight-fitting lids. When glass bottles are required for sampling, they should be coated with plastic to help prevent leakage if the glass breaks. Containers for transporting sample bottles should be compartmentalized to prevent breakage and sampling sites should be cleaned or rinsed after use.

The use of proper hygiene measures when needed can significantly reduce the risks associated with exposure to parasites or pathogens in wastewater. Most hygiene measures are simple, consume little time, and are relatively inexpensive.

REFERENCES

American Public Health Association; American Water Works Association; and Water Environment Federation (1995) *Standard Methods for the Examination of Water and Wastewater.* 19th Ed., Washington, D.C.

Ash, L.R., and Orihel, T.C. (1984) *Atlas of Human Parasitology.* American Society Clinical Pathology, Chicago, Ill.

Berk, S.G., and Gunderson, J.H. (1993) *Wastewater Organics: A Color Atlas.* Lewis Publishers, Ann Arbor, Mich.

Cable, R.M. (1977) *An Illustrated Laboratory Manual of Parasitology.* Burgess Publishing Co., Minneapolis, Minn.

Fox, J.C., et al. (1981) *Sewage Organisms: A Color Atlas.* Metropolitan Sanitation District, Chicago, Ill.

Gerardi, M.H., and Zimmerman, M.C. (1997) Parasites and Pathogens—How to Reduce Job Exposure to Disease. *Oper. Forum,* **14,** 219.

Gerardi, M.H., et al. (1983) Identification and Enumeration of Filamentous Microorganisms Responsible for Bulking Sludge. *Water Pollut. Control Assoc. Pa. Mag.,* **16,** 32.

Meyer, M.C., et al. (1988) *Essentials of Parasitology.* Wm. C. Brown Publishing, Dubuque, Iowa.

Olson, O.W. (1974) *Animal Parasites: Their Life Cycles and Ecology.* University Park Press, Baltimore, Md.

Pedersen, D.C. (1981) Density Levels of Pathogenic Organisms in Municipal Wastewater Sludge—A Literature Review. EPA-600/52-81-170, U.S. EPA, Washington, D.C.

Pritchard, M.H., and Kruse, G.O.W. (1982) *The Collection and Preservation of Animal Parasites.* University of Nebraska Press, Lincoln.

Reimers, R.S., et al. (1981) Parasites in Southern Sludges and Disinfection by Standard Sludge Treatment. EPA-600/52-81-166, U.S. EPA, Washington, D.C.

Reimers, R.S., et al. (1986) Investigation of Parasites in Sludges and Disinfection Techniques. EPA-600/51-85-022, U.S. EPA, Washington, D.C.

Schmidt, G.D., and Roberts, L.S. (1989) Foundations of Parasitology. *Times Mirror/Mosby,* Boston, Mass.

Steer, A.G., et al. (1974) A Modification of the Allen and Ridley Technique for Recovery of *Ascaris lumbricoides* Ova from Municipal Compost. *Water Res.* (G.B.), **8,** 851.

Zaman, V., and Keong, L.H. (1982) *Handbook of Medical Parasitology.* ADIS Science Press, Aust.

Chapter 12
Photomicroscopy

INTRODUCTION

Photomicroscopy is the technique of taking photomicrographs through a microscope. The equipment necessary to produce a photomicrograph (print, slide, or digital image) includes a camera, a microscope adapter, and a light source. A computer is also necessary if taking digital or analog photomicrographs. Still and video cameras are useful for recording the physical appearance of the sludge biomass, sludge bulking and foaming problems, filamentous bacteria and microscopic organisms, and effluent characteristics. Video cameras can be attached to a microscope using an adapter. Taking quality photomicrographs requires a thorough understanding of the microscope, the necessary equipment to be used, general photographic principles, and extensive practice and patience.

APPLICATIONS

Photomicrographs can be useful tools. Individually, they provide a snapshot in time of the biological condition of the wastewater treatment plant. But taken regularly, photomicrographs show useful trends key to monitoring the microlife and troubleshooting potential problems. When taking photomicrographs to initiate the trend-monitoring process, it is important to make sure that baseline photos are taken when the plant is running in good condition.

Photomicrographs can be used for a variety of purposes, including

- Indicating the conditions of the biomass and monitoring the health and age of the sludge;
- Identifying filamentous bacteria, floc structure characteristics, and zoogloea;
- Documenting the effects of chlorination during filamentous bulking or foaming problems;
- Determining total suspended solids and settleability problems;
- Documenting changes in higher life forms or floc structures when the biomass is affected from high biochemical oxygen demand loadings or toxic shocks;
- As an educational tool for training operators and laboratory personnel in the identification of wastewater organisms and their relationships to the wastewater environment;
- Preparing educational programs for the public; and
- Illustrating the diversity of the microlife present in the wastewater to appropriate audiences.

SKILL AND CARE

The goal of photomicroscopy is to record the best possible image from a microscope that is representative of the conditions of the sample. Photomicroscopy, however, can only be as good as the microscope technique used. Good microscopic techniques start with knowing the equipment and materials used. It is important to review the reference manuals for the microscope, camera, and microscope adapter to ensure that each unit is properly operated at all times. Equipment maintenance should be routinely performed following the manufacturer's recommendations. Information regarding proper maintenance and cleaning procedures for each unit can be obtained from the respective manufacturer. When not in use, the microscope and camera should be covered to ensure a dust-free environment.

EQUIPMENT FOR PHOTOMICROSCOPY

MICROSCOPE. The type of microscope used can vary from student-grade to highly technical, research-grade microscopes. The level of expertise for the person using the microscope and the desired application will help determine the best microscope to use. The most convenient microscope for photomicrographs contains an in-base illuminator (light source) and a trinocular body (Figures 12.1 and 12.2). A trinocular microscope allows the photographer to view the specimen using regular binocular eyepieces and then direct the light and image to a third tube, attached to a camera, whenever a photograph is desired. Binocular microscopes must have an eyepiece adapter attached to one of the eyepiece tubes. Extra precautions must be taken when attaching a camera to an angled eyepiece tube because additional support (such as an adjustable laboratory stand) will be needed. A special field-of-view eyepiece is used with a trinocular camera attachment. This special eyepiece differs from others in that it displays an outline of the actual area that will be included in the photomicrograph. Today, most 35-mm cameras have single-lens reflex or "through the lens" eyepieces. The image is exactly as viewed.

A stereomicroscope can be used also for larger specimens with a 35-mm camera attached also (Figure 12.3).

MICROSCOPE LENS INFORMATION. When taking photomicrographs, the size of the objective used will affect the sharpness of the image. The

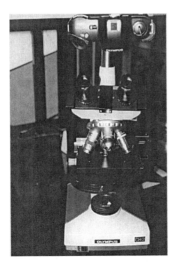 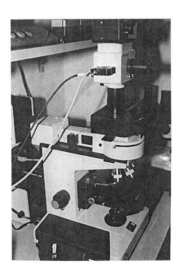

Figure 12.1 Trinocular micro-scope with 35-mm camera attached.

Figure 12.2 Research-grade trino-cular microscope with 35-mm camera attached.

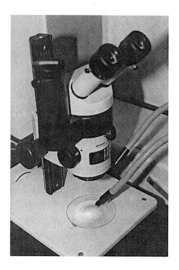

Figure 12.3   Trinocular stereomicroscope with camera attached.

condenser diaphragm controls depth of field, image contrast, and image diffraction. Typical objectives include 4X, 10X, 20X, 40X, and 100X. The following illuminations may often be used: phase contrast, bright field, and dark field. Oil immersion is used on bright field for 1000X (Figure 12.4 shows an example of a photograph taken under bright field and Figure 12.5 shows an example using phase contrast).

FILTERS. Filters can be used for polarization, diffusion, reflection, color effects, and mixed light correction. Filters can also be used for color conversion, color correction, graduated color, UV correction, or fluorescent color correction. Correction filters are typically required if the appropriate film is

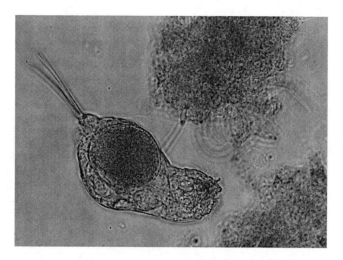

Figure 12.4   A 35-mm color negative print, 200X, bright field.

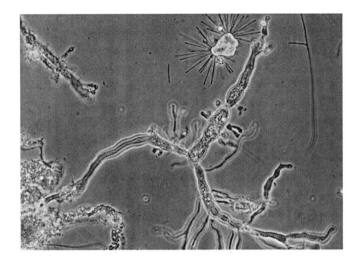

Figure 12.5   A 35-mm color negative print, 200X, phase contrast.

not available for the specific microscope light source. For precise color control, color compensating, or balancing, filters are typically needed (Table 12.1). Filters can be made of resin, glass, polyester, or gelatin. Gelatin filters require careful handling, are not easily cleaned, and are easily scratched. Glass filters are more durable and easy to clean. Newer filters, made of polyester and resin, offer many advantages over traditional gelatin filters. They are easily cleaned, substantially cheaper, offer no image degradation, and retain their color value even after extended use. Resin filters are lightweight and shatterproof.

When film speed equalization is needed because exposure would otherwise be too short, a neutral density filter may be needed to reduce image brightness. This is best done at the light source rather than the camera lens. In color photomicroscopy, the color of some stains may not appear in photomicrographs as it did in the actual image. In some cases, the use of a didymium filter can overcome this problem. This filter has the ability to enhance both blues and reds at the same time.

CAMERA. Although the microscope or camera does not need to be expensive or complicated to produce photomicrographs, it is important that, once

Table 12.1   Basic color-compensating filters.[a]

Filter color	Colors absorbed by filter
Yellow	Blue
Red	Blue and green
Green	Blue and red
Blue	Red, green, and yellow

[a] The colors of the filters vary in strength (or intensity). Beginning with the palest color available, the exposure time must be increased as the filter strength increases.

Figure 12.6   Trinocular microscope with instant camera attached.

the desired image is obtained in the microscope, it is not distorted by any photographic techniques.

Automatic shutter control is more advanced than manual shutter cable release. When using this shutter control, the shutter will open and close based on predetermined exposure times. The shutter mechanism is controlled by a cable from a control on a separate control panel. The control knob settings are based on film type and speed. This setup requires a greater understanding of exposure time, but allows exposure times adapted to the variety of the slide conditions; for example, stained versus wet-mount preparations. The operating instructions for specific camera and microscope arrangements should be obtained from the equipment manufacturer.

If the photomicrograph is needed immediately, a simple instant camera (Figure 12.6) or expensive digital and analog cameras can be used. Otherwise, a standard 35-mm camera with a built-in light meter is more practical. Comparisons of final photomicrograph quality are demonstrated in Figure 12.7. Figure 12.7 demonstrates the quality of digital versus Figures 12.4 and 12.5, which are examples using regular film and normal developing.

Film Selection. There are many film types available for photomicrography. Film-selection should be based on

- The type of camera and light source used.
- The goal of the desired output.
- The use for the photomicrographs.
- Method for storing and cataloging the photomicrographs.
- Whether black and white or color images are desired.
- Whether color film-transparency slides or color negatives are needed.

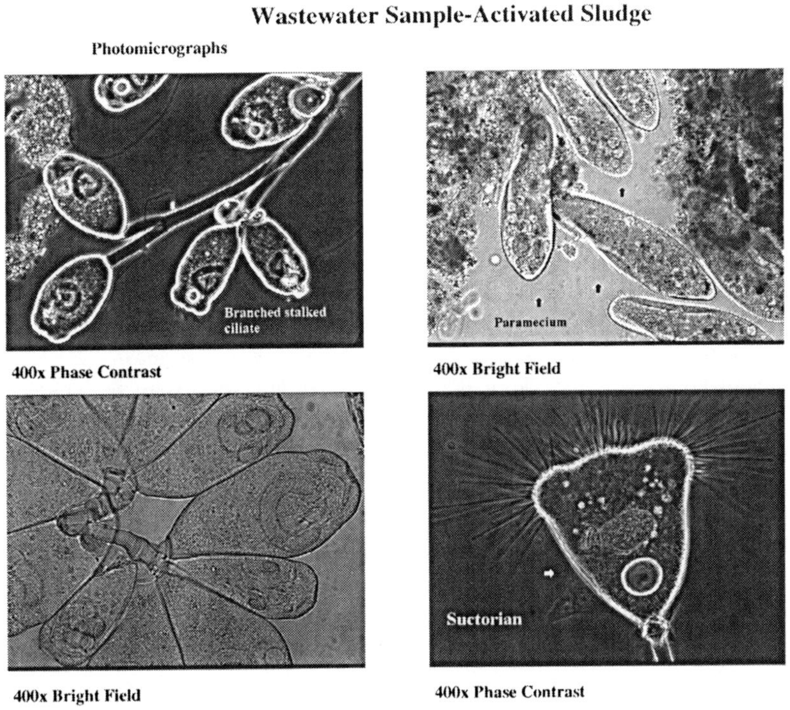

Photomicrographs

400x Phase Contrast

400x Bright Field

400x Bright Field

400x Phase Contrast

Figure 12.7   Digital photomicrographs.

The ability to reproduce the colors of the microlife is important. Thus, the film often used is 35-mm color slide film because of the convenience of handling, transporting, and filming.

Professional photographers frequently use black and white prints or slide transparencies. Slide film and black and white film are more difficult to use, but the images are sharper and contrast is slightly better than color negative prints. The developing process today for color prints is typically done by machines, which automatically adjust for slight errors or imperfections. Color negative prints are also cheaper and more easily developed, reprinted, and enlarged. Slides and black and white prints typically take longer to develop commercially.

Special 35-mm films are available for use with microscopes that have tungsten illumination. If a microscope with a tungsten light source is combined with a camera using tungsten-balanced film, then additional filters may not be needed for acceptable photomicrographs. Film sold as daylight film can be used successfully in photomicrography if the illumination has been adjusted.

Film Speed. All films are labeled with a speed rating or ISO/ASA speed. The lower the ISO/ASA number, the slower the speed of the film and more light is necessary for correct exposure. Higher speed films (400 to 1000) will require less light for proper exposure. Although use of higher speed film enables fast moving organisms to be more easily captured, the quality of the image will be grainier and fine edges fuzzier, especially when enlarged.

Different brands of films will result in minor color variations. Some brands tend to give warmer colors and certain films result in sharper, brighter, images, especially in the green color range. For images using dark field and for capturing the fast motion of higher life forms, 400 or higher speed is recommended. A local camera shop or sites on the internet can provide the characteristics of each brand and type of film.

When being processed, all films should be developed as is and without adjustment. Many photo laboratories do not know what the image is supposed to look like and may try to make corrective adjustments. Technically, no adjustments should be made on the negatives. All adjustments should be made on the cameras, lighting, and microscope long before the photo image is taken.

FACTORS AFFECTING COLOR BALANCE

Many factors may affect the color balance of photomicrographs. Filters, which transmit some colors and absorb others, can be used to facilitate the transmittance of needed color contrasts. This is more important with black and white film. Temperatures relating to film storage and temperatures at the time of exposure will affect the photomicrograph. Film should be stored at less than 10 °C (50 °F) to extend the film shelf life. Filter suggestions and temperature exposure ranges are discussed in the information sheets supplied with the film. Color corrections can also be made in the dark room.

The facility where the film is developed will affect the picture's quality. Laboratories may vary slightly on the quality of the final print. This can be caused by minor variations in the developing time, equipment, chemicals, or the developing paper used. The key to developing good photomicrographs is to find a film processor who will produce consistently high quality prints.

DETERMINING EXPOSURE

Before taking a picture, the correct exposure must be determined. This depends on the nature of the microscope, camera, the type and intensity of illumination, and numerous other factors dealing with color balancing.

On most 35-mm cameras, exposure is determined by two factors. A combination of shutter speed and f-stop (or aperture) will determine depth of field and how much light is needed.

Shutter speed determines how long the camera shutter is open. With a faster or higher number (1000 stands for 1/1000 of a second) there will be little depth of field, but fast action will be frozen. Slower shutter speeds at 1/30 and 1/15 will have greater depth of field but are not as good for freezing motion.

Aperture, or f-stop, determines how wide the camera iris is opened. Smaller numbers (f2 to f16) let in more light. Larger numbers let in less. If low light conditions exist, as when photographing at high magnifications or when the field of view is dark, the shutter speed should be reduced (slowed down) and the aperture decreased to provide more exposure time for the film. A faster shutter speed is used when high light conditions are available. Also, the faster the shutter is opened and closed, the less a moving object will blur. Minimizing movement and blurred images is difficult, but can be achieved through the use of immobilizing agents.

The appropriate exposure time can be determined only by making an exposure series or by using an exposure meter. Most automatic 35-mm cameras have these features built in. Selecting the proper exposure should be done methodically and should be fully detailed. Records must be kept of a trial run. A form used for logging information should list film type, film speed, microscope, objective, camera settings, slide or print number, light intensity setting, notes on specimen preparation, filters, magnification, and development time (if applicable). Once the slides or prints have been developed, the records can be reviewed to match the slide or print number of the optimum acceptable images to their appropriate record entries. These entry values should then be used for future photomicrography work.

EXPOSURE TIPS. Most photomicrographs tend to be underexposed. A good rule is to expose most photomicrographs approximately 50 to 100% longer than normal photographs, on average. The image viewed under the microscope almost always comes out darker on the photomicrograph. To adjust for this, use a longer exposure, larger aperture on the camera, or slower shutter speed. On some cameras, there is a correction setting or exposure compensation dial that can be adjusted for this. Typically –1 or –2 should work. The exposure compensation should be in the negative numbers when using dark field and on the positive side when viewing samples on bright field. Check individual camera manuals for variations. Different exposure compensation is also necessary depending on the ASA speed of film used.

Bracketing is a technique that also can be used on adjustable cameras. This is where the same picture is purposely under and over exposed one to two f-stops different than the suggested camera meter reading for the perfect picture to be taken.

*O*PERATING INSTRUCTIONS

The operating instructions for specific camera and microscope arrangements should be obtained from the equipment manufacturer. Following is a general procedure:

1. Be sure lenses and objectives are clean.
2. Load desired film in accordance with camera instructions. Be sure the film is compatible with the color temperature of the light (or microscope lamp).
3. Place specimen slide on the microscope stage.
4. Using the lowest power objective, scan the slide to center the field of view to be photographed.
5. View the field at greater magnification through the camera and decide which power of magnification is desired to obtain the sharpest image.
6. Adjust for proper illumination and focus.
7. If applicable, insert recommended filters (filters may simply be laid on top of the light source).
8. Recheck illumination and focus.
9. If using a system with an automatic shutter, place the shutter control mechanism between the camera body and the microscope as per the manufacturer's instructions, then properly adjust the control knob on the control panel.
10. If the camera has a built-in light meter and automatic shutter, adjust the light intensity using neutral density filters for proper exposure.
11. Press the shutter button to expose the film and advance the film.

Specimen and exposure data should be recorded in a logbook. Note that the separate shutter control may have a *lighten–darken* knob to vary exposure time. This knob should be initially set at the center position, and this setting should be recorded as part of the exposure data.

Using a log of the film speed, camera settings, microscope lenses, and other variables and comparing that log with the finished prints will prove to be a valuable educational tool while becoming familiar with the process. Practice is the only way to become technologically adept so that the photomicrographs can be obtained consistently.

*B*LACK AND WHITE PHOTOMICROGRAPHY

Although color photomicrographs add value in recording results of stained preparations and may be more aesthetically pleasing than black and white photomicrographs, black and white photomicrographs can provide good contrast and require less critical exposure.

Filters are typically needed when using black and white film. Filters are used primarily to place an organism in contrast to its environment. Maximum contrast occurs when the organism's color is completely absorbed. Filters with a deep hue have a larger effect than lighter filters of the same color. Typically, a color filter will lighten its own color on a subject and will darken a complementary color.

A yellow filter is recommended for general use in black and white photomicrography, even when contrast is not a problem. It will enhance contrast, which is especially useful when photographing wet mount preparations.

DIGITAL OR ANALOG PHOTOMICROGRAPHS

To get the best photomicrographs available today, extensive digital or analog cameras are now available. The camera is attached directly to the microscope (Figures 12.8 and 12.9). The captured digital image signals are input directly to a computer where, with different capture–archiving software packages, the images can be stored and manipulated. These images can be printed out, used in documents and presentations, sent via e-mail, enlarged or displayed, and imported to most commercially available software programs.

When using 35-mm prints, time is spent waiting to develop the film and the photographic image is not always captured correctly. With a digital or analog camera, because the exact image is seen immediately on the computer screen, no guessing or waiting time is required. Multiple shots are not required because the computer image displayed is exactly what the final image or print will be. Many images can be captured and deleted until the final image is obtained. This is helpful when taking photomicrographs of fast moving organisms. Ciliates, rotifers, and flagellates can be captured more easily.

Because the image is displayed on the computer screen, it is blown up significantly. Many fine details that are hard to see under current microscopes are easily visible with digital photomicrographs. Fine details such as hairs,

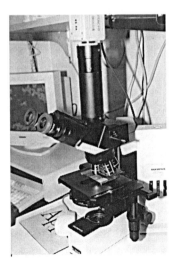

Figure 12.8 **Research-grade trinocular microscope with digital camera attached.**

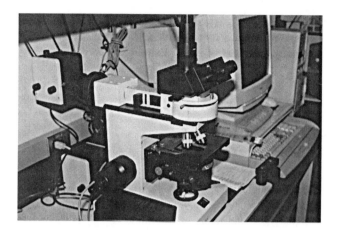

Figure 12.9 Research-grade trinocular microscope with digital camera with three-dimensional imaging attached.

moving parts, interior structures, sheaths, cell walls, and septa are visible, enabling an easier and quicker identification. For filamentous identification this is a big advantage.

Because a digital camera is already hooked up to the camera, most cameras will have a place where an additional cable can be attached to a video camera. Live, moving videos can then be recorded, if desired.

Digital cameras offer additional benefits over 35-mm prints or slides. Many software packages used in capturing and archiving the images allow manipulation of the image. Text, dates, voice, notes, scales, and measurements can be added for additional identification. This is extremely important when using photomicrographs as a monitoring tool. Certain areas of the photomicrographs can be cut, enlarged, and pasted into other images. Overlays or multiple images can also be printed out. Some software packages can capture multiple images and overlay them so that the final image has a 3-dimensional effect. Images can be printed out on a black and white, color, or photographic quality output printer.

Although digital imaging is an exceptional way to capture photomicrographs, it is not universally practical. It is more expensive to purchase a digital camera and a greater level of technical skill is required. The final goal or use for the photomicrographs will be helpful in determining the sophistication of equipment needed.

STORAGE

Film should be stored properly as film age, temperature, and storage can affect the quality of the final print. Prints, slides, and negatives also need to be stored carefully. Fingerprints, humidity, heat, and dust can ruin images.

Polypropylene sleeves are commercially available to protect against all of these.

Prints should be stored in a fashion that not only preserves the photomicrograph, but also provides easy access, especially if the images are to be used as a monitoring tool for correlation to the health of a plant's biomass. They should be cataloged, dated, and stored chronologically. System data that correlates to the plant and visual observations should be stored with the images for them to be more meaningful.

IMPROVING SPECIMEN VISIBILITY

Some organisms may need to be stained to provide structural detail that might not otherwise be clearly seen (refer to Plate 13). Stains can be used to slow organism movement to make photography easier. However, some stains can cause organisms to become distorted or lyse (burst) and cilia to bend.

Many stains are available. The Gram, Neisser, and lactophenol cotton blue stains are inexpensive and easy to use. The Gram stain is a common biological stain used primarily to differentiate between Gram-positive and Gram-negative bacteria (see Appendix A). It can also be used in the identification process to contrast the organisms against their background. Lactophenol cotton blue stain can be used to enhance fine details and sharpen images on microscopes without using phase contrast. It helps to sharpen images and bring out fine details that are not easy to see on bright field. It is concentrated and should be used in small quantities to prevent overstaining the organisms and their background. It will tend to slow down organisms. For microscopes with fluorescent photomicrography capabilities, acridine orange can be used. India ink stain can be used to quickly view polysaccharide coatings on floc structures.

The quality of the slide and cover slip used in photomicrography work also is critical. Microscope slides and cover slips should be made from high-quality glass and have the appropriate thickness. Slides approximately 1 mm thick and No.1 1-ounce cover slips (ideally ranging from 0.16 to 0.19 mm thick) should be used. They should always be free of dust and fingerprints.

Wet-mount preparations are the quickest and easiest method of viewing the microlife in their natural environment. A wastewater sample is mixed by gently inverting the sample container. Then an aliquot of the mixed sample is removed with an eyedropper or Pasteur pipette. A drop (approximately 0.05 mL) is placed on a slide and a cover slip is added. To get a sharper, clearer picture, the slide should be allowed to slightly dry out for 10 to 15 minutes. This will slow down some of the higher life forms, making a sharper picture possible. Floc structures and filamentous details will also be clearer. The slide can be prevented from drying out entirely by using a thin layer of clear nail polish around the edges.

TROUBLESHOOTING COMMON FAULTS

There are many problems that must be resolved when attempting to obtain high-quality photomicrographs of an organism or floc structure. Typical causes include low-quality optics, poor microscope techniques, and overstaining of the organism and background. Incorrect filter use may also affect the contrast or color balance. Any attempt to correct a problem should be done methodically and in a similar fashion to obtaining the first exposure determinations (with a log). Other causes of blurred images or otherwise poor output include the following:

- Vibrations caused by movements near the microscope can cause blurred pictures. Various vibration tables that can solve this problem are available commercially. If images continually come out blurred when using a 35-mm camera, a self-timer or a remote cord that will release the shutter automatically can be used to prevent camera shake.
- Dirt or dust that contaminates the lenses of the microscope, light source, or microscope slides or cover slips can make it hard to obtain a clear picture. Dirty or oily fingerprints on the slide can also cause distortions.
- A sample that is too thick on the slide can make it harder to focus clearly.
- Brownian motion as a result of water movement under the slide can make it hard to focus.
- Moving organisms can hinder obtaining a sharp focus. Because of the shallow depth of field (actual area in focus) in photomicrography, it is important to center the focus on the correct plane of interest in the field. Bright field is sometimes easier to use than phase contrast when zooming in on fine details. When an organism to be photographed is moving, an attempt should be made to *pan* or follow it while looking through the camera viewfinder, then moving the microscope slide at the same time while waiting for the specimen to slow down or stop. (This takes practice.) Using higher or faster shutter speeds on the camera helps to reduce the blurring. For those organisms moving too rapidly to photograph, movement may be slowed with an immobilizing agent such as glycerol or methyl cellulose.
- Dirt or grease buildup on glass surfaces such as the microscope slide, cover slip, or object lenses can cause haziness. These surfaces should be carefully cleaned with an appropriate cleaning solution.
- Improper adjustment of the microscope substage diaphragm or condenser can cause poor resolution.
- Dust on the cover slip, microscope slide, or other glass surfaces in the optical path can cause out of focus spots.

- When using oil-immersion lenses, dust and dirt can sometimes get in the oil, causing the image to be blurred. Another typical problem is not using enough oil on the slide to produce a sharp enough image.

CONCLUSION

Photomicrographs are an excellent tool to monitor and document the micro-life. Individually, or as part of a historical series, they can help to proactively troubleshoot potential problems. Nevertheless, this requires quality photomicrographs representative of the system. Because there are many items to consider when taking photomicrographs, it will be important to determine which equipment and methods work best for a given system.

SUGGESTED READINGS

Blaker, A.A. (1977) *Handbook for Scientific Photographs*. W.H. Freeman and Company, San Francisco, Calif.

Calumet Photographic (1996) *Essentials, The Photographer's Supply Catalog*. Bensenville, Ill.

Eastman Kodak Company (1988) *Photography through the Microscope*. 9th Ed., 1988 Bulletin Po2. Rochester, N.Y.

Polaroid Corporation (1971) How to use the Polaroid Land Instrument Camera ED-10 for Photomicrography. P589, Cambridge, Mass., 39.

Walker, M.T. (1971) *Amateur Photomicrography*. Focal Press, London, G.B., 99.

Appendix A
Test Procedures and Methods

ENUMERATION TECHNIQUES

COUNTING CHAMBER. The counting chamber, or hemacytometer, (Figure A.1) is used for performing density counts. Most chambers (A) are made so that the center platform (a), which bears the ruled counting areas, is exactly 0.1 mm lower than the raised ridges (b), which support the cover slip (c), and the platform under which the diluted sample flows. The surface of the platform is etched with a ruled area (B, e, and C). The 1-mm^2 center is subdivided in 25 squares (D).

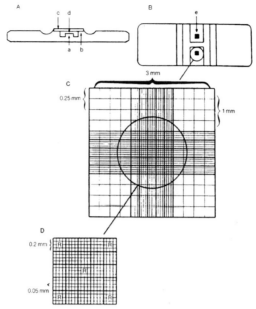

Figure A.1   Cross-section of a typical counting chamber.

To make counts, the counting area or grid is located under low power. Each counting grid is subdivided into 16 smaller squares (D). Counts can be made under 100, 400, or 1000 magnification. With each replicate aliquot, all the ova or cysts of a particular organism are counted in at least five of the counting grids (R). To remove bias in the counting procedure, counts are made in the same five grids each time. Also, to avoid counting the same object twice in any of the small squares of a grid, all cells touching the top and left of a square and none touching the lower and right boundary line are included in the count.

Because the cover slip is held at a constant height of 0.1 mm above the bottom of the chamber, the total volume (*V*) of the counting grid is

$$V = 1 \text{ mm} \times 1 \text{ mm} \times 0.1 \text{ mm}$$
$$= 0.1 \text{ mm}^3$$

The counting area is divided into 400 small squares, each with a side length (*D*) equal to 0.05 mm. The volume of the smallest square is

$$V = 0.05 \text{ mm} \times 0.05 \text{ mm} \times 0.1 \text{ mm}$$
$$= 0.000 \, 25 \text{ mm}^3$$
$$= 2.5 \times 10^{-4} \text{ mm}^3 \text{ or } 1/4000 \text{ mm}^3$$

Filament, cyst, or ova density can be expressed as

$$\text{Density} = \frac{\text{Number counter} \times (4 \times 10^3)}{\text{Number of small asquares cunted}}$$

FILAMENT MEASUREMENT. The following methods can be used to quantify filamentous microorganisms. Note, however, that these methods are detailed and time-consuming and may not be appropriate for routine mixed liquor suspended solids (MLSS) examination.

Method A (Nowak et al., 1985).

1. The activated-sludge suspension to be examined is diluted 1:1 with distilled water to minimize any false counts brought about by unequal floc distribution in a hemacytometer cell. The suspension is then mixed well.
2. Using a Pasteur pipette with the tip cut off to provide a 3-mm diameter opening, a small drop of suspension is placed on each of the two cells. The hemacytometer is then covered with its 0.5-mm cover glass.
3. At 100X total magnification (10X objective, 10X eyepiece) on the microscope, one cell of the hemacytometer is brought in focus. The filaments in the 40 diagonal squares of the grid (2 × 20 squares/diagonal) are counted. A single filament crossing one of the squares gives a count of one. For standard counting purposes, reliable results can typically be obtained with the microscope condenser set on bright field. The stage of the microscope is then moved to the second cell of the hemacytometer and the count is carried out on its diagonal squares.
4. The hemacytometer is washed with distilled water and cleaned with a piece of lint-free material. Two more drops of activated-sludge suspension are placed on the cells, and the cell counts are repeated. A total of 10 drops are used to provide a 400-square count.
5. To balance the occurrence of coiled filaments crossing a counting square, filaments crossing a diagonal square for less than a quarter of a square width are ignored, as are filaments touching the side of the square. The inside lines are used as the boundaries for those counting squares with the double line rulings.
6. For an improved Neubauer hemacytometer, the total filament length (TFL) is calculated as

$$\text{Total filament length} = \frac{\sum_{i=i}^{10} (n \times 50 \times 1000 \text{ mL}/L)}{\text{MLSS}}$$

Where

i = number of 40 square counts,
n = filament count/400 squares,
50 = hemacytometer constant,
= square width/cell volume (cm/mL), and
MLSS = mixed liquor suspended solids concentration, mg/L.

Method B (Sezgin et al., 1978).

1. Transfer 2 mL of a well-mixed activated-sludge sample of known suspended solids (SS) concentration using a wide-mouth pipette (0.8-mm diameter tip) to 1 L of distilled water in a 1.5 L beaker and stir at 95 r/min on a jar test apparatus ($G = 85$ s^{-1}) for 1 minute.
2. Using the same pipette, transfer 1.0 mL of diluted sample to a microscope counting chamber calibrated to contain 1.0 mL of sample and cover with a cover slip.
3. Using a binocular microscope at 100X magnification with an ocular micrometer scale, count the number of filaments present in the whole chamber or a known portion of it and place these in the following size classifications according to individual length: 0 to 10, 11 to 25, 26 to 50, 51 to 100, 101 to 200, 201 to 400, 401 μm, and above.
4. Express the results as

$$\text{Total extended filament length (TEFL)} = \frac{\text{TFL} \times \text{DF}}{\text{MLSS}}$$

or

$$\text{TEFL} = \text{TFL} \times \text{DF}$$

Where

TFL   = total filament length in the 1.0-mL diluted sample (μm), which equals total number filaments × average filament length;

DF   = dilution factor, which equals 500 (i.e., 1000 mL/2 mL); and

MLSS = mixed liquor suspended solids concentration (g/L), which equals 500 (i.e., 1000 mL/2 mL).

Scoring of Filament Abundance. The system shown in Table A.1 can be used to score the relative numbers of filamentous organisms present.

Table A.1   Subjective scoring of filament abundance.

Numerical Value[a]	Abundance	Explanation
0	None	
1	Few	Filaments present but only observed in an occasional floc.
2	Some	Filaments commonly observed but not present in all flocs.
3	Common	Filaments observed in all flocs but at low density (for example, 1–5 filaments per floc).
4	Very common	Filaments observed in all flocs at medium density (for example, 5–20 per floc).

Table A.1   Subjective scoring of filament abundance. (continued)

Numerical Value[a]	Abundance	Explanation
5	Abundant	Filaments observed in all flocs at high density (for example, 20 per floc).
6	Excessive	Filaments present in all flocs—appears more filaments than floc and/or filaments growing in high abundance in bulk solution.

[a] This scale from 0 to 6 represents a 100- to 1000-fold range of TEFL.

STAINING SOLUTIONS AND PROCEDURES

GRAM STAIN—MODIFIED HUCKER METHOD (APHA et al., 1992).

Solution 1. Prepare the following separately:

A		B	
Crystal violet	2 g	Ammonium oxalate	0.8 g
Ethanol, 95%	20 mL	Distilled water	80 mL

Combine A and B.

Solution 2.

Iodine	1 g
Potassium iodide	2 g
Distilled water	3000 mL

Solution 3:

Safranin 0 (2.5% in 95% ethanol)	10 mL
Distilled water	100 mL

Procedure.

1. Prepare thin smears of an activated-sludge sample on microscope slides and thoroughly air dry. Do not heat fix.
2. Stain 1 minute with solution 1; rinse 1 second with water.
3. Stain 1 minute with solution 2; rinse well with water.
4. Hold slide at an angle and decolorize with 95% ethanol added drop by drop to the smear for 25 seconds. Do not overdecolorize. Blot dry.
5. Stain with solution 3 for 1 minute; rinse well with water and blot dry.
6. Examine under oil immersion at 1000X magnification with direct illumination (not phase contrast). Blue-violet is positive; red is negative.

NEISSER STAIN.

Solution 1. Separately prepare and store the following:

A		B	
Methylene blue	0.1 g	Crystal violet	
Ethanol, 95%	5 mL	(10% w/v in 95% ethanol)	3.3 mL
Acetic acid, glacial	5 mL	Ethanol, 95%	6.7 mL
Distilled water	100 mL	Distilled water	100 mL

Mix two parts by volume of A with one part by volume of B; prepare fresh monthly.

Solution 2.

Bismark brown (1% w/v aqueous)	33.3 mL
Distilled water	66.7 mL

Procedure.

1. Prepare thin smears of an activated-sludge sample on microscope slides and thoroughly air dry. Do not heat fix.
2. Stain 30 seconds with solution 1; rinse 1 second with water.
3. Stain 1 minute with solution 2. Rinse well with water; blot dry.
4. Examine under oil immersion at 1000X magnification with direct illumination (not phase control). Blue-violet is positive (either entire cell or intracellular granules); yellow-brown is negative.

CRYSTAL VIOLET SHEATH STAIN.

Solution. Crystal violet, 0.1% w/v aqueous solution.

Procedure.

1. Mix one drop of an activated-sludge sample and one drop of crystal violet solution on a microscope slide.
2. Cover and examine at 100X magnification using phase contrast. Cells stain deep violet, while the sheaths are clear to pink.

INDIA INK REVERSE STAIN.

Solution. India ink (aqueous suspension of carbon black particles).

Procedure.

1. Mix one drop of India ink and one drop of activated-sludge sample on a microscope slide. Depending on the ink used, the sample volume may need to be reduced.
2. Place on a cover slip and observe at 1000X using phase contrast.
3. In "normal" activated sludge, the India ink particles penetrate the flocs almost completely, leaving a clear center.

4. In activated sludge containing large amounts of exocellular polymeric material, there will be large, clear areas that contain low densities of cells.

POLY–β–HYDROXYBUTYRATE STAIN.

Solution 1. Sudan black B (IV), 0.3% w/v in 60% ethanol.

Solution 2. Safranin 0 (0.5% w/v aqueous).

Procedure.

1. Prepare thin smears of an activated-sludge sample on a microscope slide and thoroughly air dry.
2. Stain 10 minutes with solution 1; add more stain if the slide starts to dry out.
3. Rinse 1 second with water.
4. Stain 10 seconds with solution 2. Rinse well with water; blot dry.
5. Examine under oil immersion at 1000X magnification with transmitted light. The polyhydroxybutyrate granules will appear as intercellular blue-black granules, while cytoplasm will be pink or clear.

SULPHUR OXIDATION ("S" TEST).

Method A.

Solution. Sodium sulfate solution ($Na_2S·9H_2O$) 1.0 g/L. Prepare weekly.

Procedure.

1. On a microscope slide, mix one drop of an activated-sludge sample and one drop of sodium sulfide solution.
2. Allow to stand open to the air for 10 to 20 minutes.
3. Place a cover slip on the preparation and gently press to exclude excess solution; remove expelled solution with a tissue.
4. Observe at 1000X magnification using phase contact. A positive "S" test gives highly refractive, yellow-colored intracellular (sulfur) granules.

This test gives variable results at times. This is because of methodological problems involving the relative concentrations of sulfide and oxygen present (sulfur oxidation is an aerobic process). An alternative procedure may be used as follows:

Method B.

Solution. Sodium thiosulfate solution ($Na_2S_2O_3·5H_2O$) 1 g/100 mL.

Procedure.

1. Allow an activated-sludge sample to settle, and transfer 20 mL of clear supernatant to a 100-mL Erlenmeyer flask.
2. Add 1 to 2 mL of activated sludge to the flask.
3. Add 1 mL of thiosulfate solution to the flask (final thiosulfate concentration is 2 m*M*).
4. Shake the flask overnight at room temperature.
5. Observe a sample of the shaken activated sludge at 1000X magnification using phase contrast.

METHYLENE BLUE STAIN.

Solution.

Methylene blue	0.01 g
Absolute alcohol	100 mL

Procedure.

1. Using dilute quantities of the stain, add a drop at the edge of the cover slip and allow it to seep under the cover slip. Do not overstain.
2. For sludge, add one drop of methylene blue to one drop of sludge and cover with a cover slip.

LUGOL'S IODINE STAIN.

Solution.

Potassium iodide	10 g
Iodine crystals	5 g
Distilled H_2O	100 mL

Mix reagents and filter into brown bottle. Place a stopper tightly on the bottle and store away from light.

PRESERVATIVES AND FIXATIVES

MERTHIOLATE–IODINE–FORMALIN PRESERVATIVE.

Solution (MIF stock solution).

Distilled water	50 mL
Formalin (commercial)	5 mL
Tincture of merthiolate (1:1000)	40 mL
Glycerol	1 mL

Add 9.4 mL of MIF stock to 0.6 mL of Lugol's iodine stain to make up merthiolate–iodine–formalin preservative. Mix well and store in brown bottle.

BURROW'S POLYVINYL ALCOHOL FIXATIVE. The polyvinyl alcohol (PVA) fixative solution is made from modified Schandinn's fixative and PVA mixture, which is prepared separately and later mixed.

Solution (Modified Schandinn's fixative).

Mercuric chloride crystals	4.5 g
Ethanol	21.0 mL
Glacial acetic acid	5.0 mL

Dissolve the mercuric chloride in the alcohol in a stoppered flask (50 or 125 mL) by swirling the mixture in the flask at intervals. Add the acetic acid and stopper, and mix by swirling.

Solution (PVA) mixture.

Glycerol	1.4 mL
PVA powder, pretested	31.0 mL
Distilled water	62.5 mL

1. In a small beaker, add the glycerol to the PVA powder and mix thoroughly with a glass rod until all particles appear coated with glycerol. Scrape the mixture into a 125-mL flask.
2. Add the distilled water and stopper, and leave at room temperature for 3 hours to overnight. Swirl mixture occasionally to mix.

Procedure (PVA fixative solution).

1. Prepare a water bath (or large beaker of water) of 70 to 75 °C (94 to 103 °F).
2. Place the loosely stoppered flask containing the PVA mixture in the bath for 10 minutes, swirling frequently.
3. When the PVA powder appears to be mostly dissolved, pour in the Schandinn's fixative solution, reapply the stopper, and swirl the mixture.
4. Continue to swirl the mixture in the bath for 2 to 3 minutes to dissolve the remainder of the PVA, allow bubbles to escape, and clear the solution.
5. Remove the flask from the water bath and let cool. Store PVA fixative solution in a screw-cap or glass-stoppered bottle.

MISCELLANEOUS REAGENTS. Formalin. Commercial formalin is approximately a 40% solution of formaldehyde gas in water. This solution is known as 100% formalin and is diluted accordingly when concentrations of formalin are desired. For example, 5% formalin, which is 2% formaldehyde, contains 95 mL of distilled water and 5 mL of commercial neutral formalin.

Sheather's Sucrose Solution. Mix 600 g of sucrose (granulated sugar), 320 mL of distilled water, and 6.5 g of phenol (liquor or crystal). Heat the distilled water to dissolve the sugar, then add the phenol. The specific gravity of the solution is 1.40.

Zinc Sulfate Solution. Mix 330 g of zinc sulfate ($ZnSO_4$) and 1000 mL of distilled water. Heat to dissolve the zinc sulfate and add water to adjust to a specific gravity of 1.8 or 1.2.

DILUTED SLUDGE VOLUME INDEX PROCEDURE (Stobbe, 1964).

1. Set up several 10-L graduated cylinders (the number will depend on prior knowledge of the settleability of the sludge).
2. Using well-clarified secondary effluent, prepare a series of twofold dilutions of the activated sludge (i.e., no dilution, 1:1 dilution, 1:3 dilution).
3. Stir the graduated cylinders individually for 30 to 60 seconds using a plunger to resuspend and uniformly distribute the sludge solids.
4. Allow the activated sludge to settle for 30 minutes.
5. Observe the 30-minute settled sludge volume (SV_{30}) in the graduated cylinder where the settled volume is less than or closest to 200 mL ($SV_{30} \leq 200$ mL).
6. Calculate the diluted sludge volume index.

$$\text{Diluted sludge volume index} = \frac{2^n}{SV_{30} \times SS}$$

Where
SV_{30} = 30-minute settled sludge volume (mL/L),
n = number of twofold dilutions required to obtain $SV_{30} \leq 200$ mL, and
SS = suspended solids concentration of the undiluted activated sludge (g/L).

REFERENCES

American Public Health Association; American Water Works Association; and Water Pollution Control Federation (1992) *Standard Methods for the Examination of Water and Wastewater.* 18th Ed., Washington, D.C.

Nowak, G., et al. (1985) Effects of Feed Pattern and Dissolved Oxygen on Growth of Filamentous Bacteria. Paper presented at 58th Annu. Water Pollut. Control Fed. Conf., Kansas City, Mo.

Sezgin, M., et al. (1978) A Unified Theory of Filamentous Activated Sludge Bulking. *J. Water Pollut. Control Fed.,* **50**, 362.

Stobbe, G. (1964) Uber das Verhalten von belebtem Schlamn aufsteigender Wasserbewegung. Veroffentlichingen Inst. Siedlingwasserwritschaft Techischen Hochschule Hannover, Heft, 18.

Glossary

acclimation The dynamic response of a system to the addition or deletion of a substance, until equilibrium is reached; adjustment to a change in the environment.

acetic acid CH_3COOH, a clear, colorless liquid or crystalline mass with a pungent odor, miscible with water or alcohol; crystallizes in deliquescent needles; a component of vinegar. Also known as ethanoic acid.

acetogenesis Microbial conversion of carbon dioxide to sugars and acetate.

acidophyle (acidophile) An organism that grows best under acid conditions (down to a pH of 1).

actinomycete Highly filamentous bacteria; moldlike bacteria.

activated sludge Sludge withdrawn from a secondary clarifier following the activated-sludge process; consists mostly of biomass with some inorganic settable solids. Return sludge is recycled to the head of the process; waste (excess) sludge is removed for conditioning.

active transport The pumping of ions or other substances across a cell membrane against an osmotic gradient, that is, from a lower to higher concentration.

acute Having a sudden onset, sharp rise, and short course.

adaptation The occurrence of genetic changes in a population or species as the result of natural selection so that it adjusts to new or altered environmental conditions.

adhesion Intimate sticking together of surfaces.

adsorption The adherence of a material to the surface of a solid.

adsorption coefficient A measure of a material's tendency to absorb to soil or other particles.

aeration The addition of oxygen to water or wastewater, usually by mechanical means, to increase dissolved oxygen levels and maintain aerobic conditions.

aerobe An organism that requires air or free oxygen to maintain its life processes.

aerobic Requiring, or not destroyed by, the presence of free elemental oxygen.

aerobic oxidation (respiration) The conversion of organics such as carbohydrates to bacterial cells, carbon dioxide, water, and energy.

aerosol A suspension of colloidal particles in air or another gas.

aerotolerant anaerobe Microbes that grow under both aerobic and anaerobic conditions but do not shift from one mode of metabolism to another as conditions change. They obtain energy exclusively by fermentation.

agar A gelatinous substance extracted from red algae; commonly used as a medium for laboratory cultivation of bacteria.

agglutinate To unite or combine into a group or mass.

aggregate Crowded or massed into a dense cluster.

alcohol Any of a class of organic compounds containing the hydroxyl group –OH.

aldehyde One of a class of organic compounds containing the CHO radical.

algae Photosynthetic microscopic plants, which, in excess, can contribute taste and odor to potable water and deplete dissolved oxygen on decomposition.

aliphatic Of or pertaining to any organic compound of hydrogen and carbon characterized by a straight chain of the carbon atoms; the subgroups of such compounds are alkanes, alkenes, and alkynes.

alkaline Having a pH greater than 7.

alkalinity The alkali concentration or alkaline quality of an alkali-containing substance.

alkane A member of a series of saturated aliphatic hydrocarbons having the empirical formula C_nH_{2n+2}.

alkene One of a class of unsaturated aliphatic hydrocarbons containing one or more carbon-to-carbon double bonds.

alkyne One of a group of organic compounds containing a carbon-to-carbon triple bond.

aloricate Without a lorica (a hard protective shell or case).

ambient The environmental conditions in a given area.

amination A process in which the amino group ($-NH_2$) is introduced to organic molecules.

amino acid Any of the organic compounds that contain one or more basic amino groups and one or more acidic carboxyl groups and that are polymerized to form peptides and proteins; only 20 of the more than 80 amino acids found in nature serve as building blocks for proteins; examples are tyrosine and lysine.

amoeba Protozoa that use pseudopodia (temporary extensions of the cell) for locomotion and feeding.

amorphous Pertaining to a solid that is noncrystalline, having neither definite form nor structure.

anabiosis State of suspended animation induced by desiccation and reversed by addition of moisture.

anabolic Pertaining to the part of metabolism involving the union of smaller molecules into larger molecules; the method of synthesis of tissue structure.

anabolism A part of metabolism involving the union of smaller molecules into larger molecules; the method of synthesis of tissue structure.

anaerobic (1) A condition in which no free oxygen is available. (2) Requiring, or not destroyed by, the absence of air or free oxygen.

anaerobic respiration Respiration under anaerobic conditions. The terminal electron acceptor, instead of oxygen in the case of regular respiration, can be CO_2, Fe^{2+}, fumarate, nitrate, nitrite, nitrous oxide, sulfur, sulfate, etc. Note that anaerobic respiration still uses an electron transport chain to dump the electron while fermentation does not.

anaerobiosis A mode of life carried on in the absence of molecular oxygen.

---

anatomical Relating to the structure of organisms.

anion An ion that is negatively charged.

annelid A multisegmented worm.

anoxic Lacking in oxygen.

anoxygenic See anaerobic.

antagonistic Mutual opposition; nullification by one substance of a chemical or physical effect due to another.

anterior Situated near or toward the front or head of an animal body.

anthrone test A test to determine total carbohydrates; the reagent hydrolyzes polysaccharides to monosaccharides and forms a colored compound in the presence of monosaccharides.

anthropogenic Referring to environmental alterations resulting from the presence or activities of humans.

antibiotic A substance produced by a microorganism, which, in dilute solution, inhibits or kills another microorganism.

antibody Any of the body of globulins (proteins) that combine specifically with antigens and neutralize toxins.

anus The posterior orifice of the alimentary canal.

aperture An opening, for example, the opening in a photographic lens that admits light.

aquaculture Cultivation of natural faunal resources of water.

aqueous Relating to or made with water.

aromatic Pertaining to or characterized by the presence of at least one benzene ring.

artificial support media Material such as a glass slide or polyurethane foam that is introduced to the wastewater treatment system to allow colonization by protozoa for later collection and examination.

asexual reproduction Reproduction (cell division, spore formation, fission, or budding) without union of individuals or germ cells.

assimilation Conversion of nutritive materials to protoplasm.

asthma A pulmonary disease marked by labored breathing, wheezing, and coughing; cause may be emotional stress, chemical irritation, or exposure to an allergen.

Aufwuchs community A community of microscopic plants and animals associated with the surfaces of submerged objects.

autoclavable Able to be sterilized by autoclaving.

autolysin An enzyme that causes the cell that made it to self-destruct.

autotroph An organism capable of synthesizing organic nutrients directly from simple inorganic substances, such as carbon dioxide and inorganic nitrogen.

bacilloid Rod shaped.

bacteria A group of universally distributed, rigid, essentially unicellular microscopic organisms lacking chlorophyll. They perform a variety of biological treatment processes, including biological oxidation, sludge digestion, nitrification, and denitrification.

bacterivorous Feeding on bacteria.

baseline A sample used as comparative reference point when conducting further tests or calculations.

basophyle (basophile) An organism that prefers, or can tolerate, alkaline conditions, typically in the pH range of 8 to 11.

bench-scale testing A small-scale test or study used to determine whether a technology is suitable for a particular application.

benthic Relating to the bottom or bottom environment of a body of water.

beta (β) oxidation The process of fatty-acid catabolism in which two-carbon fragments are removed in succession from the carboxyl end of the chain.

binary fission Form of asexual reproduction in some microbes in which the parent organism splits into two independent organisms.

binocular eyepiece Device that divides a beam of light into two different beams and directs them through two eyepiece tubes so that both eyes can be used at the microscope.

bioactivation Transformation of a chemical within an organism to a biochemically active metabolite.

bioaugmentation The addition of commercially prepared cultures of saprophytic or nitrifying bacteria.

bioavailability The degree to which an agent, such as a drug or nutrient, becomes available at the physiological site of activity.

biochemical oxygen demand (BOD) A standard measure of wastewater strength that quantifies the oxygen consumed in a stated period of time, usually 5 days, and at 20 °C.

biodegradability The characteristic of a substance to be broken down by organisms.

biodegradation The breaking down of a substance by organisms.

bioenergetics The branch of biology dealing with energy transformations in living organisms.

biofilm An accumulation of microbial growth.

bioindicator Species or group of species that is representative and typical for a specific status of an ecosystem, that appears frequently enough to serve for monitoring, and whose population shows a sensitive response to changes.

biological transformation A chemical transformation within a living cell.

biomass The dry weight of living matter, including stored food, present in a species population and expressed in terms of a given area or volume of the habitat.

biomass carriers An inert material on which organisms can grow while feeding on wastewater organics and other materials.

biooxidation The process by which living organisms, in the presence of oxygen, convert organic matter to a more stable or a mineral form. *(The process by which all living things obtain energy for metabolic processes.)*

biorecalcitrant Resistant to biological treatment.

biosynthesis Production, by synthesis or degradation, of a chemical compound by a living organism.

bivalve shell Having a shell composed of two valves.

blank sample Laboratory simulated test material known to be free of the chemical being analyzed. A portion of the blank material is used to test the method, apparatus, and reagents for interferences or contamination.

bog Soft, waterlogged ground; marsh.

bottomland Low-lying land along a river.

brackish (1) Of water, having salinity values ranging from approximately 0.50 to 17.00 parts per thousand. (2) Of water, having less salt than sea water, but undrinkable.

bristleworm Aquatic annelid of the class Polychaeta that has both errant and sedentary species.

Brownian motion Random movements of small particles suspended in a fluid, caused by the statistical pressure fluctuations over the particle.

buccal opening Cavity of the mouth.

buffering capacity The relative ability of a buffer solution to resist pH change with addition of an acid or base.

bulking Clouds of billowing sludge that occur throughout secondary clarifiers and sludge thickeners when the sludge does not settle properly. In the activated sludge process, bulking is usually caused by filamentous bacteria or bound water.

bulking sludge Low-density activated sludge that settles poorly.

butyric acid $CH_3CH_2CH_2COOH$, a colorless, combustible liquid with a boiling point of 163.5 °C; soluble in water, alcohol, and ether; used in the synthesis of flavors, in pharmaceuticals, and in emulsifying agents.

calcareous Containing calcium.

carapace A bony or chitinous case or shield covering the back or part of the back of an animal.

carbohydrates Any of the group of organic compounds composed of carbon, hydrogen, and oxygen, including sugars, starches, and celluloses.

carbon fixation A process occurring in photosynthesis whereby atmospheric carbon dioxide gas is combined with hydrogen obtained from water molecules.

---

carbonaceous biochemical oxygen demand The portion of biochemical oxygen demand whereby oxygen consumption is caused by oxidation of carbon; usually measured after a sample has been incubated for 5 days.

carboxylic acid Any of a family of organic acids characterized by the presence of one or more carboxyl groups.

carcinogen Any agent that incites development of a carcinoma or any other sort of malignancy.

carrier protein A protein that transports specific substances through the cell membrane in which it is embedded and to the cell. Different carrier proteins are required to transport different substances because each one is designed to recognize only one substance or group of similar substances.

catabolic Pertaining to the metabolic change of complex to simple molecules.

catabolism That part of metabolism concerned with the breakdown of large protoplasmic molecules and tissues, often with the liberation of energy.

catalyze To modify the rate of a chemical reaction as a catalyst.

cation An ion that is positively charged.

caudal Toward, belonging to, or pertaining to the tail or posterior end.

causative Effective or operating as a cause or agent.

caustic Alkaline or basic.

cavitation Pitting of a solid surface such as metal or concrete.

cellulolytic Of, pertaining to, or causing the hydrolysis of cellulose.

cellulose The main polysaccharide in living plants, forming the skeletal structure of the plant cell wall; a polymer of β-D-glucose units linked together, with the elimination of water for form chains composed of 2000 to 4000 units.

centrate The liquid remaining after solids have been removed in a centrifuge.

cephalothorax The body division comprising the united head and thorax of arachnids and higher crustaceans.

chain-of-custody Documentation of times, dates, and personnel involved in sample collection, transport, and analysis.

chelating agent An organic compound in which atoms form more than one coordinate bond with metals in solution.

chelation A chemical process involving formation of a heterocyclic ring compound that contains at least one metal cation or hydrogen ion in the ring.

chemisorption A chemical adsorption process in which weak chemical bonds are formed between gas or liquid molecules and a solid surface.

chemoautotroph Any of a number of autotrophic bacteria and protozoans that do not carry out photosynthesis.

chemoheterotroph An organism that derives energy and carbon from the oxidation of preformed organic compounds.

chemolithotroph An organism that obtains its energy from the oxidation of inorganic compounds.

chemoorganotroph An organism that requires an organic source of carbon and metabolic energy.

chemotroph An organism that extracts energy from organic and inorganic oxidation–reduction reactions.

chitin A white or colorless amorphous polysaccharide that forms a base for the hard outer integuments of crustaceans, insects, and other invertebrates.

chlorination The addition of chlorine to water or wastewater, usually for the purpose of disinfection.

chlorine Cl_2, an oxidant commonly used as a disinfectant in water and wastewater treatment.

chlorophyll The generic name for one of several plant pigments that function as photoreceptors of light energy for photosynthesis.

chloroplast A plastid that contains chlorophyll and is the site of photosynthesis.

cholesterol A sterol produced by all vertebrate cells, particularly in the liver, skin, and intestine, and found most abundantly in nerve tissue.

chromatography A method of separating and analyzing mixtures of chemical substances by chromatographic adsorption.
    elution—the removal of absorbed species from a porous bed or chromatographic column by means of a stream of liquid or gas.

gas—a separation technique involving passage of a gaseous moving phase through a column containing a fixed absorbent phase; used principally as a quantitative analytical technique for volatile compounds.

high performance liquid—a laboratory technique, a type of column chromatography that uses a combination of several separation techniques to separate substances at a high resolution.

thin-layer—chromatography using a thin layer of powdered medium on an inert sheet to support the stationary phase.

chronic Long continued; of long duration.

cilia Relatively short centriole-based, hairlike processes on certain anatomical cells and motile organisms.

ciliate Any of various protozoans of the phylum Ciliophora having cilia at some stage in their life cycle.

clarifier A quiescent tank used to remove suspended solids by gravity settling. Also called sedimentation or settling basins, they are usually equipped with a motor-driven chain-and-flight or rake mechanism to collect settled sludge and move it to a final removal point.

cleavage The state of being split or cleft; a fissure or division.

cloaca The chamber that functions as a respiratory, excretory, and reproductive duct in certain invertebrates.

coagulant Chemical added to destabilize, aggregate, and bind together colloids and emulsions to improve settleability, filterability, or drainability.

coccoid (cocci) A spherical bacterial cell.

coenzyme The nonprotein portion of an enzyme; a prosthetic group that functions as an acceptor of electrons or functional groups.

coliform bacteria Rod-shaped bacteria living in the intestines of humans and other warm-blooded animals.

colloid Suspended solid with a diameter smaller than 1 m that cannot be removed by sedimentation alone.

colloidal material Finely divided solids that will not settle but may be removed by coagulation, biochemical action, or membrane filtration.

colonization The establishment of immigrant species in a peripherally unsuitable ecological area; occasional gene exchange with the parental population occurs, but generally the colony evolves in relative isolation and in time may form a distinct unit.

co-metabolism The metabolic transformation of a substance while a second substance serves as primary energy or carbon source.

commensal An organism living in a state of *commensalism (an interspecific, symbiotic relationship in which two different species are associated, wherein one is benefitted and the other is neither benefitted nor harmed).*

community Aggregation of organisms characterized by a distinctive combination of two or more ecologically related species; an example is a deciduous forest.

compactability A measure of the ability of a substance to be compacted, which increases bulk density and decreases porosity.

competition The inter- or intraspecific interaction resulting when several individuals share an environmental necessity.

competitive exclusion The result of a competition in which one species is forced out of part of the available habitat by a more efficient species.

competitive exclusion principle (Gause's principle) A statement that two species cannot occupy the same niche simultaneously.

complexation See complexing.

complexing Formation of a complex compound.

composting A managed method for the biological decomposition of organic materials frequently used to stabilize sludge.

compound microscope Microscope consisting of an objective and eyepiece mounted in a drawtube.

concentrator A plant in which materials are concentrated.

condenser A lens or mirror used to shorten or diminish in size.

confluence (1) A stream formed from the flowing of two or more streams. (2) The place where such streams form.

confocal microscopy A system of (usually) epifluorescent light microscopy in which a fine laser beam of light is scanned over the object through the objective lens. The technique is particularly good at rejecting light from

---

outside the plane of focus and thus produces higher effective resolution than is normally achieved.

conjugation A process involving contact between two bacterial or ciliate cells during which genetic material is passed from one cell to the other.

conserve To protect from loss or depletion.

constitutive Making a thing what it is; essential.

contractile Having the power to shorten or diminish in size.

contractile vacuole A tiny, intracellular membranous bladder that functions in maintaining intra- and extracellular osmotic pressures in equilibrium as well as excretion of water, such as occurs in protozoans.

coprozoic Living in fecal matter.

copulation The sexual union of two individual organisms, resulting in insemination or deposition of the male gametes in close proximity to the female gametes.

corona The upper portion of a body part (as a tooth or the skull).

corrosion Gradual destruction of a metal or alloy due to chemical processes such as oxidation or the action of a chemical agent.

coryneform (1) Of bacteria, rod shaped with one end substantially thicker than the other. (2) Any Gram-positive, nonmotile bacteria that occur as irregularly shaped rods and resemble members of the genus *Corynebacterium*.

covalent bond A bond in which each atom of a bond pair contributes one electron to form a pair of electrons.

crustaceans Any of a large class of mostly aquatic mandibulate arthropods that have a chitinous or calcerous and chitinous exoskeleton, a pair of often modified appendages on each segment, and two pairs of antennae and that include lobsters, shrimps, crabs, wood lice, water fleas, and barnacles.

cryptobiosis (cryptobiotic state) A state in which the metabolic rate of an organism is reduced to an imperceptible level.

cuticle An external layer usually secreted by epidermal cells.

cyanobacteria Blue-green algae.

cycle Process in which the system returns to the original point at the end of a complete operation, for example, nitrogen cycle in plant life.

cyrtophorid An ostracod crustacean larva with a long first pair of antennae and lost swimming capability in the second pair. Also, a group of ciliate protozoa.

cyst Resting stage formed by some bacteria, nematodes, and protozoa in which the whole cell is surrounded by a protective layer, not the same as endospore.

cytomembrane The internal membrane found in nitrifying bacteria in biofilm.

cytopharyngeal A region of the plasma membrane or cytoplasm of some ciliated and flagellated protists specialized for endocytosis; a permanent oral canal.

deanimate Removal from a molecule of the amino group.

decarboxylate To remove the carboxyl radical, especially from amino acids and protein.

dechlorinating agent A chemical added to a solution to neutralize or bind excess chloride.

definitive host The host in which a parasite reproduces sexually.

degradation The breakdown of substances by biological action.

deleterious Having a harmful effect, injurious.

denitrification The reduction of nitrate or nitrite to gaseous products such as nitrogen, nitrous oxide, and nitric oxide; brought about by denitrifying bacteria.

depolymerize Decomposition of macromolecular compounds into relatively simple compounds.

derivatization The conversion of a chemical compound to a derivative.

desiccation Thorough removal of water from a substance, often with the use of a desiccant.

detention time The theoretical time required to displace the contents of a tank or unit at a given rate of discharge.

detritus Disintegrated matter.

diabetes Any of various abnormal conditions characterized by excessive urinary output, thirst, and hunger.

diaphragm Any opening in an optical system that controls the cross section of a beam of light passing through it to control light intensity, reduce aberration, or increase depth of focus. Also known as lens stop.

diatom Unicellular, microscopic algae with a boxlike structure consisting principally of silica.

dichotomous keys Scheme for making identifications by asking a series of questions that can be answered directly and without quantification, for example, motile, rod shaped, spore produced?

didymium A mixture of rare-earth elements made up chiefly of neodymium and praseodymium and used especially for coloring glass for optical fibers.

diffraction A modification light undergoes in passing by the edges of opaque bodies or through narrow slits or in being reflected from ruled surfaces and in which the rays appear to be deflected and produce fringes of parallel light and dark-colored bands.

diffusion The spatial equalization of one material throughout another.

diffusion coefficient The weight of a material in grams diffusing across an area of 1 cm^2 in 1 second in a unit concentration gradient. Also known as diffusivity.

diffusivity See diffusion coefficient.

digester A tank or vessel used for sludge digestion.

digestion The process of converting food to an absorbable form by breaking it down to simpler chemical compounds.

dilute To make thinner or less concentrated by adding a liquid such as water; to lessen the force, strength, purity, or brilliance of, especially by admixture.

dilution rate The reciprocal of hydraulic retention time.

dinoflagellate Any of a class (Dinoflagellatea) of unicellular flagellate protozoa that include luminescent forms, forms important in marine food chains, and forms causing red tide.

dipeptide A peptide that yields two molecules of amino acid on hydrolysis.

diphtheria An acute febrile contagious disease marked by the formation of a false membrane especially in the throat and caused by a bacterium (*Corynebacterium diphtheriae*) that produces a toxin causing inflammation of the heart and nervous system.

disaccharide Any of the class of compound sugars that yield two monosaccharide units on hydrolysis.

discoid Being flat and circular in form.

disinfection With a disinfectant, the killing or reduction in numbers of waterborne fecal and pathogenic bacteria and viruses in potable water supplies or wastewater effluents.

dispersion Scattering and mixing.

dissimilation A form of microbial metabolism, oxidative in nature; breakdown.

dissimilative sulfate reduction Generation of sulfate from hydrogen sulfide.

dissolved oxygen The oxygen dissolved in a liquid.

distilled water Water that has been freed of dissolved or suspended solids and organisms by distillation.

dominance The influence that a controlling organism has on numerical composition or internal energy dynamics in a community.

dry mass The mass of organisms calculated after removal from the media and drying of the culture.

dysentery Inflammation of the intestine characterized by pain, intense diarrhea, and the passage of mucus and blood.

ecosystem A functional system that includes the organisms of a natural community together with their environment.

ectoplasm The outer relatively rigid granule-free layer of the cytoplasm usually held to be a gel reversibly convertible to a sol.

effluent Partially or completely treated water or wastewater flowing out of a basin or treatment plant.

electrolyte A chemical compound that, when molten or dissolved in certain solvents (usually water), will conduct an electric current.

---

electron acceptor An atom or part of a molecule joined by a covalent bond to an electron donor.

electron donor An atom or part of a molecule that supplies both electrons of a duplet forming a covalent bond.

electron microscope A microscope in which a beam of electrons focused by means of electromagnetic lenses is used to produce an enlarged image of a minute object on a fluorescent screen or photographic plate.

electron transport system The components of the final sequence of reactions in biological oxidations; composed of a series of oxidizing agents arranged in order of increasing strength and terminating in oxygen.

electrophile An electron-deficient ion or molecule that takes part in an electrophilic process.

electrostatic properties Pertaining to electricity at rest, such as an electric charge on an object.

elimination To get rid of, remove; to excrete as waste.

eluant A liquid used to extract one material from another, as in chromatography.

elute To extract one material from another, usually by means of a solvent.

embryonic state A state resembling or having characteristics of an embryo.

emulsion A stable dispersion of one liquid in a second immiscible liquid.

encephalitis Inflammation of the brain.

encyst The process of forming or becoming enclosed in a cyst or capsule.

endemic Restricted or native to a particular locality or region.

endocarditis Inflammation of the endocardium.

endoplasm The inner relatively fluid part of the cytoplasm (the protoplasm) of a cell external to the nuclear membrane.

endospore Differentiated cell formed within the cells of certain Gram-positive bacteria and extremely resistant to heat and other harmful agents.

enrichment A process that changes the isotopic ratio in a material.

enteric Of or within the intestine.

enterococci A group of sphere-shaped bacteria that normally inhibit the intestines of humans or other animals.

enteropathogenic Tending to produce disease in the intestinal tract.

enumeration To determine the number of, count.

enzyme Any of a group of catalytic proteins that is produced by living cells and that mediates and promotes the chemical processes of life without itself being altered or destroyed.

epidemic A disease that occurs simultaneously in a large fraction of the community.

epidemiological Pertaining to the study of the incidence, distribution, and control of disease in a population.

esophagus The tubular portion of the alimentary canal between the pharynx and the stomach.

ester The compound formed by the elimination of water and the bonding of an alcohol and an organic acid.

estuarine Of, relating to, or formed in an estuary.

eubacteria A superclassification (above kingdom level) of all prokaryotes, excludes Archaebacteria.

euglenoid Any member of the division Euglenophyta.

eukaryote A cell with a definitive nucleus.

eutrophication Nutrient enrichment of a lake or other water body, typically characterized by increased growth of planktonic algae and rooted plants. It can be accelerated by wastewater discharges and polluted runoff.

execretory system Those organs concerned with solid, fluid, or gaseous excretion.

exoenzyme An enzyme that functions outside the cell in which it is synthesized.

exogeneous Due to an external cause; not arising within the organism.

exoskeleton The external supportive covering of certain invertebrates, such as arthropods.

exotoxin A soluble poisonous substance given off during the growth of an organism.

exposure (1) The act or fact of exposing or being exposed. (2) The product of the duration and intensity of light striking a photosensitive material; the former is controlled by shutter speed, the latter by aperture.

extracellular organic binding substance Insoluble polysaccharides.

extracellular polymers Insoluble "slime" of polysaccharides.

extreme thermophile (hyperthermophile) An organism living at temperatures near 110 °C.

eyespot A simple light-sensitive organ of pigment or pigmented cells covering a sensory termination.

facilitated diffusion The movement of polar molecules across a cell membrane via protein transporters.

facultative anaerobe A microorganism that grows equally well under aerobic and anaerobic conditions.

fauna Animals.

fermentation Changes in organic matter or organic wastes brought about by anaerobic microorganisms and leading to the formation of carbon dioxide, organic acids, or other simple products.

fibril A small filament or fiber.

field-of-view eyepiece A special eyepiece equipped with an indicator for the actual area being photographed.

filamentous microorganisms Bacteria and fungi that grow in threadlike colonies resulting in a biological mass that will not settle and may interfere with drainage through a filter.

filiform Threadlike or filamentous.

flagella Relatively long, whiplike, centriole-based locomotor organelles on some motile cells.

flagellate An organism that propels itself by means of a relatively long, whiplike, centriole-based locomotor organelle; a flagella.

flatworm The common name for members of the phylum Platyhelminthes; individuals are dorsoventrally flattened.

floc Collections of smaller particles agglomerated into larger, more easily settleable particles through chemical, physical, or biological treatment.

flocculation In water and wastewater treatment, the agglomeration of colloidal and finely divided suspended matter after coagulation by gentle stirring by either mechanical or hydraulic means. In biological wastewater treatment when coagulation is not used, agglomeration may be accomplished biologically.

fluorescent Exhibiting or capable of the emission of electromagnetic radiation, especially light.

food chain The scheme of feeding relationships by trophic levels, which unites the member species of a biological community.

food-to-microorganism ratio (F:M) In the activated-sludge process, the loading rate expressed as the amount of biochemical oxygen demand per amount of mixed liquor suspended solids per day.

food vacuole A membrane-bound organelle in which digestion occurs in cells capable of phagocytosis.

food web A modified food chain that expresses feeding relationships at various, changing trophic levels.

formic acid HCOOH, a colorless, pungent, toxic, corrosive liquid melting at 8.4 °C; soluble in water, ether, and alcohol; used as a chemical intermediate and solvent in dyeing and electroplating processes and in fumigants. Also known as methanoic acid.

free-living Living or moving independently.

free radical An atom or a diatomic or polyatomic molecule that possesses at least one unpaired electron.

fretting Surface damage usually in an air environment between two surfaces, one or both of which are metals in close contact under pressure and subject to a slight relative motion.

f-stop An aperture setting for a camera lens; indicated by the f-number.

fulvic acids A group of organic acids formed in water on degradation of organic plant matter.

fungi Nucleated, usually filamentous, spore-bearing organisms devoid of chlorophyll.

gastroenteritis An inflammation of the stomach and intestinal tract.

gastrotrich Any of a phylum (Gastrotricha) of minute aquatic pseudocoelomate animals that usually have a spiny or scaly cuticle and cilia on the ventral surface.

gelatin A protein derived from the skin, white connective tissue, and bones of animals.

gene The basic unit of inheritance.

genus (plural genera) A taxonomic category that includes groups of closely related species; the principal subdivision of a family.

geochemistry The study of the chemical composition of the various phases of the earth and the physical and chemical processes that have produced the observed distribution of elements and nuclides in these phases.

geosmin A metabolite of certain blue-green algae and actinomycetes that causes an earthy–musty taste and odor in water supplies.

glucose $C_6H_{12}O_6$, a monosaccharide; occurs free or combined and is the most common sugar.

glycocalyx The outer component of a cell surface, outside the plasmalemma; usually contains strongly acidic sugars, hence it carries a negative electric charge.

glycolysis The enzymatic breakdown of glucose or other carbohydrate, with the formation of lactic acid or pyruvic acid and the release of energy in the form of adenosinetriphosphate.

glycoprotein A conjugated protein in which the nonprotein group is a carbohydrate.

gonidia A sexual reproductive cell or groups of cells arising in a special organ on or in a gametophyte.

graduated cylinder A tall narrow container with a volume scale used especially for measuring liquids.

graininess Mottled (spotted) appearance of an exposed photograph.

Gram-negative bacteria Bacteria that do not retain the purple dye used in the Gram staining method due to a higher lipid content of the cell wall.

Gram-positive bacteria Bacteria that retain the purple dye used in the Gram staining method.

Gram stain A common staining procedure used to differentiate bacteria into Gram-negative and Gram-positive categories.

granule A small grain or pellet; particle.

graticule A scale on transparent material in the focal plane of an optical instrument for locating and measuring objects.

gravity separation A process in which the components of a nonhomogeneous mixture separate themselves by the force of gravity in accordance with their individual densities.

group translocation A process of actively importing compounds to the bacterial cell. The compound diffuses into the cell passively and is immediately modified (for example, by phosphorylation) so that it cannot diffuse back out.

growth factors Any factor, genetic or extrinsic, that affects growth.

growth rate Increase in the number of bacteria in a population per unit time.

gymnostome Any of a class of ciliates (Gymnostomatea) characterized by having simple or inconspicuous oral cilia and usually with extrusomes for food capture.

habitat The part of the physical environment in which a plant or animal lives.

halobacteria Bacteria that live in conditions of high salinity.

halogen Any of the elements of the halogen family consisting of fluorine, chlorine, bromine, iodine, and astatine.

heavy metal A metal that can be precipitated by hydrogen sulfide in an acid solution and that may be toxic to humans in excess of certain concentrations.

helminth Any parasitic worm.

hemacytometer A specifically designed, ruled, and calibrated glass slide used with a microscope to count organisms.

---

hemicellulose A type of polysaccharide found in plant cell walls in association with cellulose and lignin; soluble in and extractable by dilute alkaline solutions.

Henry's law The mass of a gas dissolved by a given volume of liquid at a constant temperature is proportional to the pressure of the gas.

hepatitis An acute viral disease that results in liver inflammation and may be transmitted by direct contamination of the water supply by wastewater.

hermaphroditic Having the sex organs and many of the secondary sex characteristics of both male and female.

heterogeneous Consisting of or involving dissimilar elements or parts; not homogeneous.

heterotroph An organism that obtains nourishment from the ingestion and breakdown of organic matter.

heterotrophic bacteria A type of bacteria that derives its cell carbon from organic carbon; most pathogenic bacteria are heterotrophic bacteria.

homogeneous Pertaining to a substance having uniform composition or structure.

homogenize To make homogeneous.

hopper A funnel-shaped receptacle with an opening at the top for loading and a discharge opening at the bottom for bulk delivery of material.

humic substances (humus) Substances pertaining to or derived from humus.

humus Total of the organic compounds in soil exclusive of undecayed plant and animal tissues, their "partial decomposition" products, and the soil biomass. The term is often used synonymously with soil organic matter.

hydraulic retention time (HRT) (hydraulic residence time) The average time of retention of liquid in the treatment system. Vessel volume divided by the rate of liquid removed resulting in a unit of time.

hydraulic shear Force exerted by water flowing past a surface that works to remove objects adhered to that surface.

hydric Pertaining to, characterized by, or requiring considerable moisture.

hydride A compound containing hydrogen and another element.

hydrocarbon An organic compound consisting predominantly of carbon and hydrogen.

hydrogen peroxide H_2O_2, unstable, colorless, heavy liquid boiling at 158 °C; soluble in water and alcohol; used as a bleach, chemical intermediate, rocket fuel, and antiseptic.

hydrology The science that treats the occurrence, circulation, distribution, and properties of the waters of the earth and their reactions with the environment.

hydrolysis (1) The reaction of a solute with water in aqueous solution. (2) A change in the chemical composition of matter produced by combination with water. Sometimes loosely applied in wastewater practice to the liquefaction of solid matter in a tank as a result of biochemical activity. (3) Usually a chemical degradation of organic matter; also the combination of water with metallic cations; also used in reference to the amount of anionic groups in certain flocculants.

hydrolyze To split complex molecules into simpler molecules by chemical reaction with water.

hydrophilic Having an affinity for, attracting, adsorbing, or absorbing water.

hydrophobic Lacking an affinity for, repelling, or failing to adsorb or absorb water.

hydrophytic Plants that require large amounts of water for growth.

hygienic practices Practices tending to promote or preserve health.

hymenostome Any of a subclass of ciliates in the class Oligohymenophorea having fairly uniform somatic ciliation and specialized oral cilia within a buccal cavity.

hyphae The microfilaments composing the mycelium of a fungus.

hypochlorite OCl⁻, chlorine anion commonly used as an alternative to chlorine gas for disinfection.

hypotrich A subclass of ciliates in the class Spirotrichea. Typically with a dorsoventrally flattened body with cirri (compound ciliary organelles) on the ventral surface and with conspicuous oral cilia (rows of membranelles).

illuminator A device for producing, concentrating, or reflecting light.

immersion Placement into or within a fluid, usually water.

immunization Natural or artificial development of resistance to a specific disease.

immunocompromised Having the immune system impaired or weakened.

in situ In the original location.

incineration The process of reducing the volume of a solid by the burning of organic matter.

indicator species A species whose requirements most reflect those of the species community in the habitat of concern, usually used to indicate habitat quality and predict future conditions.

inert Lacking an activity, reactivity, or effect.

influent Water or wastewater flowing into a basin or treatment plant.

ingestion The act or process of taking food and other substances into the animal or protozoan body.

inhibition The act of repressing or restraining a physical or chemical reaction.

inorganic Pertaining to or composed of chemical compounds that do not contain carbon as the principle element (excepting carbonates, cyanides, and cyanates), that is, matter other than plant or animal.

interior Of, relating to, or located in the inside; inner.

intermediate A precursor to a desired product.

intermediate host The host in which a parasite multiplies asexually.

intracellular Within a cell.

ion exchange A chemical reaction in which mobile hydrated ions of a solid are exchanged, equivalent for equivalent, for ions of like charge in solution.

ionize To convert totally or partially to ions.

iris diaphragm An adjustable diaphragm of thin opaque plates that can be turned by a ring to regulate the aperture of a lens.

ISO/ASA speed The speed rating for a particular film, for example, a film having an ISO rating of 25 is a slow film and needs a lot of exposure.

isomerization A process by which a compound is changed into an isomer, for example, conversion of butane into isobutane.

isothermal Having constant temperature.

jar test A test procedure using laboratory glassware for evaluating coagulation, flocculation, and sedimentation in a series of parallel comparisons.

juvenile Physiologically immature or undeveloped; young.

ketone One of a class of chemical compounds of the general formula RR'CO where R and R' are alkyl, aryl, or heterocyclic radicals; the groups R and R' may be the same or different or incorporated to a ring. The ketones, acetone, and methyl ethyl ketone are used as solvents, and ketones in general are important intermediates in the synthesis of organic compounds.

kineties Rows of cilia in ciliates.

kinetoplastid A class of small flagellate protozoa in the phylum Euglenozoa; some are pathogenic parasites to humans and other vertebrates.

larva The early form that hatches from the egg of many insects, alters chiefly in size while passing through several molts, and is finally transformed into a pupa or chrysalis from which the adult emerges.

leachate Fluid that percolates through solid materials or wastes and contains suspended or dissolved materials or products of the solids.

lecithin Any of a group of phospholipids having the general composition $CH_2OR_1 \cong CHOR_2 \cong CH_2OPO_2OHR_3$ in which R_1 and R_2 are fatty acids and R_3 is choline and having emulsifying, wetting, and antioxidant properties.

ligand The molecule, ion, or group bound to the central atom in a chelate or a coordination compound.

lignin A substance that, with cellulose, forms the woody cell walls of plants and cements them together.

lime The term usually refers to ground limestone (calcium carbonate), hydrate lime (calcium hydroxide), or burned lime (calcium oxide).

lipids One of a class of compounds that contain long-chain aliphatic hydrocarbons and their derivatives, such as fatty acids, alcohols, amines, amino alcohols, and aldehydes; includes waxes, fats, and derived compounds.

lipolytic Capable of hydrolyzing fats, oils, or waxes.

lipopolysaccharide Any of a class of conjugated polysaccharides consisting of a polysaccharide combined with a lipid.

lipoprotein Any of a class of conjugated proteins consisting of a protein combined with a lipid.

locomotion The act of moving or the ability to move from place to place.

longitudinal Pertaining to the lengthwise dimension.

loricate Having a lorica (a hard protective case or shell).

lyophilization Rapid freezing of a material, especially biological specimens for preservation, at a very low temperature, followed by rapid dehydration by sublimation in a high vacuum.

lyse (breakup) To undergo *lysis* (the rupture of a cell that results in loss of its contents).

macroinvertebrate Any larger-than-microscopic animal that has no backbone or spinal column.

macromolecule A large molecule in which there is a large number of one or several relatively simple structural units, each consisting of several atoms bonded together.

macronucleus A relatively large densely staining nucleus that is believed to exert a controlling influence over the trophic (nutritional) activities of most ciliated protozoa.

macronutrients Elements, such as potassium and nitrogen, essential in large quantities for cellular growth.

macroorganisms Any animal or plant life that is larger than microscopic.

macrophytic Pertaining to a macroscopic plant in an aquatic environment.

magnification The apparent enlargement of an object by an optical instrument.

malicious Given to, marked by, or arising from malice.

malodorous Having a bad odor; foul-smelling.

manifold A pipe fitting with several lateral outlets for connecting one pipe to another.

marsh An area of low-lying wet land.

mass spectrometry An analytical technique for identification of chemical structures, determination of mixtures, and quantitative elemental analysis, based on application of the mass spectrometer.

mastax The muscular pharynx in rotifers.

mean cell residence time (MCRT) Average time that a given unit of cell mass stays in the activated sludge biological reactor (aeration tank). It is usually calculated as the ratio of total mixed liquor suspended solids in the reactor to that of wastage.

mesophile Bacteria that grow best at temperatures between 25 and 40 °C.

mesophilic That group of bacteria that grow best within the temperature range of 25 to 40 °C.

mesosaprobic Moderately to heavily polluted.

metabolic Of or resulting from metabolism.

metabolism The physical and chemical processes by which foodstuffs are synthesized into complex elements, complex substances are transformed into simple ones, and energy is made available for use by an organism.

metabolize To transform by metabolism; to subject to metabolism.

metalization The coating of a metal or nonmetal surface with a metal, as by metal spraying or vacuum evaporation.

metaphosphate The inorganic anion PO_3^- or a compound containing it.

metazoa Micro- and macroscopic multicellular organisms with cells organized in layers or groups as specialized tissues or organ systems.

methane formers (methanogens) Group of anaerobic bacteria responsible for conversion of organic acids to methane gas and carbon dioxide.

methanogenesis The biological production of methane.

methemoglobinemia A pathological condition caused by chemical interference with the oxygen-transfer mechanism of the blood. It may be caused in infants by drinking water high in nitrates.

microaerophilic Pertaining to those microorganisms requiring free oxygen but in very low concentration for optimum growth.

microbial Pertaining to microorganisms too small to be seen with the naked eye.

microbial community All microbial organisms within an environment.

microbial ecology The study of the interactions of microorganisms with their physical environment and each other.

microcarriers See biomass carriers.

microflora Microscopic plants.

microfungus Fungus with a microscopic fruiting body (organ specialized for producing spores).

microhabitat A small, specialized, and effectively isolated location.

micrometer An instrument used with a microscope to measure minute distances.

micronucleus A minute nucleus concerned with reproductive and genetic functions in most ciliated protozoans.

micronutrients Nutrients, such as magnesium and calcium, required in minute quantities for cellular growth.

microorganism Microscopic organism, either plant or animal, invisible or barely visible to the naked eye. Examples are algae, bacteria, fungi, protozoa, and viruses.

microtubules Any of the minute cylindrical structures that are widely distributed in the protoplasm and are made up of longitudinal fibrils.

midgut The mesodermal intermediate part of an invertebrate intestine.

mineralization The conversion of an organic material to an inorganic form by microbial decomposition.

mitochondria Any of various round or long cellular organelles of most eukaryotes that are found outside the nucleus; produce energy for the cell through cellular respiration; and are rich in fats, proteins, and enzymes.

mixed liquor suspended solids (MLSS) The concentration of suspended solids in activated sludge mixed liquor, expressed in milligrams per liter.

mixotroph An organism able to assimilate organic compounds as carbon sources while using inorganic compounds as electron donors.

molt To cast off an outer covering periodically.

monomer A simple molecule that is capable of combining with a number of like or unlike molecules to form a polymer; a repeating structure unit within a polymer.

monosaccharide A carbohydrate that cannot be hydrolyzed to a simpler carbohydrate; a polyhedric alcohol having reducing properties associated with an actual or potential aldehyde or ketone group; classified on the basis of the number of carbon atoms as triose (3C), tetrose (4C), pentose (5C), and so on.

morphology The structure or form of an organism or feature of an organism.

most probable number Statistical analysis technique based on the number of positive and negative results during testing of multiple portions of equal volume.

motile Being capable of spontaneous movement.

mucous Of, relating to, or resembling mucus.

mucus A viscid slippery substance secreted by the mucous membranes that line body passages and cavities that communicate directly or indirectly with the exterior.

mutagen An agent that rises the rate of mutation above the spontaneous rate.

mutation An abrupt change in the genotype of an organism, not resulting from recombination; genetic material may undergo qualitative or quantitative alteration or rearrangement.

mutualism A necessary and beneficial interaction between two organisms living in the same environment.

mycelial Relating to the mass of interwoven filamentous hyphae that form the vegetative portion on the thallus of a fungus and are often submerged in another body (as of soil, organic matter, or the tissues of a host).

mycolic acids Saturated fatty acids found in the cell walls of mycobacteria, *Nocardia*, and corynebacteria. Chain lengths can be as great as 80, and the mycolic acids are found in waxes and glycolipids.

mycoplasma Any of the genus (*Mycoplasma* of the family Mycoplasmataceae) of pleomorphic Gram-negative chiefly nonmotile bacteria that are mostly parasitic—usually in mammals; called pleuropneumonia-like organism.

---

nassulids Any ciliate of the class Nassophorea.

Neisser stain A staining process used to identify bacteria.

nematode Any member of a group of unsegmented worms.

nematophagus Organisms, such as fungi, that feed on nematodes or nonsegmented roundworms.

neutrophile An organism that grows best under neutral conditions (pH of 7).

niche The role of an organism within an environment.

nitrification Formation of nitrous and nitric acids or salts by oxidation of the nitrogen in ammonia; specifically, oxidation of ammonium salts to nitrites and oxidation of nitrites to nitrates by nitrifying bacteria.

nitrogenous biochemical oxygen demand The portion of biochemical oxygen demand whereby oxygen consumption results from the oxidation of nitrogenous material; measured after the carbonaceous oxygen demand has been satisfied.

nitrosamine Yellow, aromatic, organic compound.

nocardioforms A group of actinomycetes that form mycelia that break up into rod-shaped or coccoid elements. Certain genera in this group are pathogenic to man.

n-octanol–water partition coefficient The equilibrium ratio, commonly expressed as a logarithm, of the concentration of a solute in two immiscible solvents, one less polar than the other, that is, octanol and water.

nonmotile Not capable of spontaneous movement.

non-spore-forming bacteria Bacteria that do not form endospores under adverse environmental conditions.

nucleophile A species possessing one or more electron-rich sites, such as an unshared pair of electrons or the negative end of a polar bond.

nucleus A small mass of differentiated protoplasm rich in nucleoproteins and surrounded by a membrane; found in most animal and plant cells; contains chromosomes; and functions in metabolism, growth, and reproduction.

nutrient deficiency Lack of substances that are assimilated by an organism and promote growth; generally applied to nitrogen and phosphorus in wastewater but also to other essential and trace elements.

---

nymphs Any of various immature insects, especially a larva of an insect (as a grasshopper, true bug, or mayfly) with incomplete metamorphosis that differs from the imago, especially in size and in its incompletely developed wings and genitalia.

objective lens (objective) A lens that forms an image of an object.

obligate aerobes Bacteria that can survive only in the presence of dissolved oxygen.

obligate anaerobes Bacteria that can survive only in the absence of dissolved oxygen.

ocular micrometer A scale in the field of vision of an eyepiece used as a measuring device.

ogliosaprobic Slightly polluted.

organelle A specialized cellular part that is analogous to an organ.

organic See organic matter.

organic matter Chemical substances of animal or vegetable origin, or more correctly, containing carbon and hydrogen.
    particulate—material of plant or animal origin that is suspended in fluid.
    soluble—capable of being dissolved in a fluid.

organic substrate Carbon- and hydrogen-containing compounds that are amenable to biological degradation.

orthophosphate (1) A salt that contains phosphorus as $(PO_4)^{-3}$. (2) Product of hydrolysis of condensed (polymeric) phosphates. (3) A nutrient required for plant and animal growth.

osmotic Of or relating to osmosis (the passage of a pure solvent through a semipermeable membrane).

osmotroph An organism that can absorb nutrients through the cell surface, as is commonly the case with bacteria and fungi.

oxic Presence of air or oxygen

oxidant Compound that gives up oxygen easily, removes hydrogen from another compound, or attracts negative electrons. Also known as an oxidizing agent.

oxidation (1) A chemical reaction in which the oxidation number (valence) of an element increases due to the loss of one or more electrons by that element. Oxidation of one element is accompanied by simultaneous reduction of the other reactant. (2) The conversion of organic materials to simpler, more stable forms with the release of energy, accomplished by chemical or biological means. (3) The addition of oxygen to a compound.

oxidation–reduction potential The potential required to transfer electrons from an oxidant to a reductant that indicates the relative strength potential of an oxidation–reduction reaction.

oxygenase An oxidoreductase that catalyzes the indirect incorporation of oxygen to its substrate.

oxygenated Treated, infused, or combined with oxygen.

oxygenic See aerobic.

ozone O_3, a strong oxidizing agent with disinfection properties similar to chlorine; also used in odor control and sludge processing.

parasite An organism living in or on another organism.

parthenogenetic Pertaining to reproduction in which offspring are produced by an unfertilized female; usually used when the species in question normally reproduces sexually.

particulate Matter in the form of small liquid or solid particles.

partition To divide into parts, pieces, or sections.

passive transport See facilitated diffusion.

Pasteur pipette An ungraduated glass pipette, typically used to transfer small volumes of liquid sample to a microscope slide.

pasteurization The application of heat for a specified time to a liquid food or beverage to enhance its keeping properties by destroying harmful microorganisms.

pathogen A disease-producing agent; usually refers to living organisms.

pathogenic organisms Organisms that cause disease in the host organism by their parasitic growth.

pellet A small, solid, or densely packed ball or mass.

peptidoglycan A polymer that is found in bacterial cell walls.

peritonitis Inflammation of the peritoneum.

peritrich A bell-shaped or tubular microorganism of the class Peritria, characterized by a wide oral opening surrounded by cilia.

permeability The ability of a membrane or other material to permit a substance to pass through it.

permeable Having a texture that permits water to move through perceptibly.

permease Any of a group of enzymes that mediate the phenomenon of active transport.

persistence The tendency of a refractory, nonbiodegradable material to remain essentially unchanged after being introduced to the environment.

pH The reciprocal of the logarithm of the hydrogen ion concentration in gram moles per liter. On the 0 to 14 scale, a value of 7 at 25 °C (77 °F) represents a neutral condition. Decreasing values indicate increasing hydrogen ion concentration (acidity), and increasing values indicate decreasing hydrogen ion concentration (alkalinity).

phagocytosis (1) The process by which a cell is engulfed and broken down by another for purposes of defense or sustenance. (2) The uptake of extracellular materials by the formation of a pocket from the cellular membrane and its subsequent pinching off.

phagotroph An organism that ingests nutrients by phagocytosis.

pharyngeal Relating to, located, or produced in the region of the pharynx.

pharyngitis Inflammation of the pharynx.

phase-contrast microscope A microscope that translates differences in the phase of light transmitted through or reflected by the object into differences of intensity in the image.

phenol An organic pollutant known as carbolic acid occurring in industrial wastes from petroleum-processing and coal-coking operations.

phosphatase Enzyme that hydrolyzes phosphoric acid esters of carbohydrates.

phospholipids Any of a class of esters of phosphoric acid containing one or two molecules of fatty acid, an alcohol, and a nitrogenous base.

---

phosphorylate To change (an organic substance) into an organic phosphate.

photoautotroph An organism that derives energy from light and manufactures its own food.

photoheterotroph An organism using light as a source of energy and organic materials as a carbon source.

photolysis The use of radiant energy to produce chemical changes.

photomicroscopy Photographic microscopy.

photophosphorylation Phosphorylation induced by light energy in photosynthesis.

photoreceptor A highly specialized, light-sensitive cell or group of cells containing photopigments.

photosynthesis The process of converting carbon dioxide and water to carbohydrates, activated by sunlight in the presence of chlorophyll.

phototroph An organism that utilizes light as a source of metabolic energy.

phycocyanin Any of the bluish-green protein pigments in the cells of blue-green algae.

phycoerythrin Any of the red protein pigments in the cells of red algae.

physiochemical Pertaining to physiology and chemistry.

physiological In accord with or characteristic of the normal functioning of a living organism.

phytoplankton Plankton consisting of plants, such as algae.

phytotoxic Toxic to plants.

pilus (plural pili) Any filamentous appendage other than flagella on certain Gram-negative bacteria.

pin floc Small floc particles that agglomerate poorly.

pinpoint floc See pin floc.

plaque A clear area representing a colony of viruses on a plate culture formed by lysis of the host cell.

plasmid An extrachromosomal genetic element found in certain bacteria.

plasmolysis Shrinking of the cytoplasm away from the cell wall due to exosmosis by immersion of a plant cell in a solution of higher osmotic activity.

plug flow Conditions in which fluid and fluid particles pass through a tank and are discharged in the same sequence that they enter.

pneumatic controller A valve in which compressed air forced against a diaphragm is opposed by a spring to regulate fluid flow.

pneumonia An acute or chronic inflammation of the lungs caused by numerous microbial, immunological, physical, or chemical agents.

polaritization The process of producing a relative displacement of positive and negative bound charges in a body by applying an electric field.

polarity Quality of the electrical charge, whether it is positive or negative, on particles or electrodes.

polio An infectious viral disease occurring mainly in children and, in its acute forms, attacking the central nervous system and producing paralysis, muscular atrophy, and often deformity.

poly-β-hydroxybutyrate A common organic carbon storage material of prokaryotic cells consisting of a polymer of beta hydroxybutyrate or other beta-alkanoic acids.

polychaete A small worm common in seas and estuaries and often chosen for bioassays of coastal regions.

polyelectrolyte A compound consisting of a chain of organic molecules used as coagulants or coagulant aids. See also polymer.

polymer (1) Substance made of giant molecules formed by the union of simple molecules (monomers). (2) Common term for polyelectrolyte.

polymerization The bonding of two or more monomers to produce a polymer.

polypeptide A chain of amino acids linked together by peptide bonds but with a lower molecular weight than a protein; obtained by synthesis or by partial hydrolysis of protein.

polyphosphate Inorganic compound in which two or more phosphorus atoms are joined together by oxygen.

---

polypropylene A crystalline, thermoplastic resin made by the polymerization of propylene, C_3H_6; the product is tough and hard, resists moisture, oils, and solvents, and withstands temperatures up to 170 °C; used to make molded articles, fibers, film, rope, printing plates, and toys.

polysaccharide A carbohydrate composed of many monosaccharides.

polysaprophic Very heavily polluted.

polyurethane foam A solid or spongy cellular material produced by the reaction of a polyester (such as glycerin) with a diisocyanate (such as toluene diisocyanate) while carbon dioxide is liberated by the reaction of a carboxyl with the isocyanate; used for thermal insulation, soundproofing, and padding.

population dynamics The aggregate of processes that determine the size and composition of any population.

posterior The hind end of an organism.

precipitation The process of producing a separable solid phase within a liquid medium; represents the formation of a new condensed phase, such as vapor or gas condensing to liquid droplets; a new solid phase gradually precipitates within a solid alloy as a result of slow, inner chemical reaction; in analytical chemistry, precipitation is used to separate a solid phase in an aqueous solution.

predator–prey relationship
    predators—animals that prey on other animals as a source of food.
    prey—animals consumed by other animals (predators).

primary producers Organisms that convert the energy of the sun to carbon-based energy.

product A substance produced by a chemical change.

prokaryotic Lacking a distinct nucleus.

proliferation The growth or production by multiplication of parts, as in budding or cell division.

propagation The process of increasing in number.

propionic acid CH_3CH_2COOH, water and alcohol soluble, clear, colorless liquid with pungent aroma; boils at 140 °C; used to manufacture various propionates in nickel-electroplating solutions for perfume esters and artificial flavors, for pharmaceuticals, and as a cellulsics solvent. Also known as methylacetic acid; propanoic acid.

---

prosthecate bacteria Single-celled microorganisms that differ from typical unicellular bacteria in having one or more appendages that extend from the cell surface.

prostomium The portion of the head anterior to the mouth in annelids and mollusks.

protein Any of a class of high-molecular-weight polymer compounds composed of a variety of a-amino acids joined by peptide linkages.

proteinaceous Pertaining to any material having a protein base.

proteolytic Enzymes that hydrolyze proteins.

protoplast The living portion of a cell considered as a unit; includes the cytoplasm, the nucleus, and the plasma membrane.

protozoa Predominantly single-celled, phagotrophic, eukaryotic organisms, including amoebae, ciliates, and flagellates.

pseudopodia Temporary projections of the protoplast of amoeboid cells in which cytoplasm streams actively during extension and withdrawal.

psychrophile An organism that thrives at low temperatures.

pupa An intermediate, usually quiescent, stage of a metamorphic insect (as a bee, moth, or beetle) that occurs between the larva and the imago, is usually enclosed in a cocoon or protective covering, and undergoes internal changes by which larval structures are replaced by those typical of the imago.

putrefaction Decomposition of organic matter, particularly the anaerobic breakdown of proteins by bacteria, with the production of foul-smelling compounds.

pyridine C_5H_5N, organic base; flammable, toxic yellowish liquid, with penetrating aroma and burning taste; soluble in water, alcohol, ether, benzene, and fatty oils; boils at 116 °C; used as an alcohol denaturant, solvent, in paints, medicine, and textile dyeing.

pyrolyze To break apart complex molecules into simpler units by the use of heat.

qualitative Of, pertaining to, or concerning quality.

quantitative (1) Of, relating to, or expressing in terms of quantity. (2) Of, relating to, or involving the measurement of quantity or amount.

---

quiescent Inactive or still, dormant.

random number tables A specially constructed table of numbers used in statistics to obtain a representative sample population, that is, each member of the population has an equal chance of being included in the sample.

recalcitrant Not responsive to treatment.

reducing power Electrons stored in reduced electron carriers such as NADH, NADPH, and FADH$_2$.

reduction Chemical reaction in which an element gains an electron.

refractory organic Organic substances that are difficult or impossible to metabolize in a biological system.

remedial action Something done to correct or improve a deficiency.

residence time The period of time that a volume of liquid remains in a reactor or system.

resin Any of a class of solid or semisolid organic products of natural or synthetic origin with no definite melting point, generally of high molecular weight; most resins are polymers.

resolution The process or capability of making distinguishable the individual parts of an object, closely adjacent optical images, or sources of light.

respiration Intake of oxygen and discharge of carbon dioxide as a result of biological oxidation.

return activated sludge (RAS) Settled activated sludge that is returned to mix with raw or primary settled wastewater.

rhizosphere Thin film aerobic region subject to the influence of plant roots and characterized by a zone of increased microbiological activity.

rosette A structure or marking resembling a rose.

rostrum A beak or beaklike projection.

rotating biological contactor (RBC) A fixed-growth biological treatment device whereby organisms are grown on circular disks mounted on a horizontal shaft that slowly rotates through wastewater.

rotifers Minute, multicelled aquatic animals.

saccharolytic Capable of breaking down sugars.

saprobic system Method of evaluating the overall degree of organic pollution on a single scale, using the presence or absence of individual species or community structures as parameters.

saprophyte An organism that lives on decaying organic matter.

saprozoic Feeding on decaying organic matter; applied to animals.

Sarcina A bacterial genus.

scavenger An organism that feeds on carrion, refuse, and similar matter.

scum Floatable materials found on the surface of primary and secondary settling tanks consisting of food wastes, grease, fats, paper, foam, and similar materials.

sedentary Permanently attached.

Sedgewick–Rafter method A method for the quantitative determination of microscopic organisms (i.e., those larger than bacteria) in water.

sedimentation The process of subsidence and decomposition of suspended matter carried by water, wastewater, or other liquids by gravity. It is usually accomplished by reducing the velocity of the liquid below the point at which it can transport the suspended material. Also called settling. May be enhanced by coagulation or flocculation.

segmented worm See annelid.

septic Putrid, rotten, foul smelling; anaerobic.

septum (plural septa) A partition or dividing wall between two cavities.

sessile Permanently attached to the substrate.

seta A slender, usually rigid bristle or hair.

settleability The tendency of suspended solids to settle.

sexual reproduction Reproduction involving the paired union of special cells from two organisms.

sheath A protective case or cover.

shock load A sudden increase in hydraulic or organic loading to a treatment plant.

shutter A camera attachment that exposes the film or plate by opening an aperture.

siliceous Of, relating to, or containing silica or a silicate.

single lens reflex A camera having a single lens that forms an image that is reflected to the viewfinder or recorded on film.

sinusitis Inflammation of a paranasal sinus.

sloughing The disattachment of accumulated biological solids from trickling filter media.

sludge volume index (SVI) The ratio of the volume in milliliters of sludge settled from a 1000-mL sample in 30 minutes to the concentration of mixed liquor in milligrams per liter multiplied by 1000.

sludge wasting Removal of solids, including biomass, from wastewater treatment processes.

solids retention time (SRT) The average time of retention of suspended solids in a biological waste treatment system, equal to the total weight of suspended solids leaving the system per unit of time (usually per day).

solubility The amount of a substance that can dissolve in a solution under a given set of conditions.

soluble Capable of being dissolved.

solvent That part of a solution that is present in the largest amount, or the compound that is normally liquid in the pure state (as for solutions of solids or gases in liquids).

sorption A general term used to encompass the processes of adsorption, absorption, desorption, ion exchange, ion exclusion, ion retardation, chemisorption, and dialysis.

sorption partition coefficient The concentration of a chemical adsorbed to the matrix divided by the concentration in solution.

species A taxonomic category ranking immediately below a genus and including closely related, morphologically similar individuals that actually or potentially interbreed.

spicule A minute, slender, pointed (usually hard) body.

spirillar Corkscrew shaped.

spirochaetes Any of an order (Spirochaetales) of slender spirally undulating bacteria, including those causing syphilis and relapsing fever.

spirotrich An order of ciliated protozoans having membranelles around the mouth and few cilia elsewhere on the body.

spontaneous Occurring without application of an external agency because of the inherent properties of an object.

sporozoa Also known as the Apicomplexa, is a phylum of Protozoa, typically producing spores during the asexual stages of the life cycle.

stabilization Maintenance at a relatively nonfluctuating level, quantity, flow, or condition.

stage micrometer A finely divided scale ruled on a microscope slide and used to calibrate the threads or lines across the field of view.

starch Any one of a group of carbohydrates or polysaccharides of the general composition $(C_6H_{10}O_5)_x$, occurring as organized or structural granules of varying size and markings in many plant cells; it hydrolyzes to several forms of dextrin and glucose; its chemical structure is not completely known, but the granules consist of concentric shells containing at least two fractions: an inner portion called amylose and an outer portion called amylopectin.

stereoscopic binocular microscope (stereomicroscope) An assembly of two microscopes into a single binocular microscope to give a stereoscopic (three-dimensional) view and a correct rather than an inverted image.

sterilization The destruction or removal of all living organisms within a system.

stirred sludge volume index (SSVI) Volume occupied by the solids after the settling period.

storage granule (1) Membrane-bound vesicles containing condensed secretory materials (often in an inactive zymogen form). (2) Granules found in plastids or in cytoplasm, assumed to be food reserves; often of glycogen or other carbohydrate polymer.

stramenopiles Classification of protists named for the tripartite tubular hairs associated with most members of the group; contains the zooflagellates, amoebae, funguslike organisms, sporozoan-like organisms, and varied forms of algae.

stratification The arrangement of a body of water into two or more horizontal layers of differing characteristics, especially densities.

streptococcal infection Infection with bacteria of the genus *Streptococcus*.

strict aerobes See obligate aerobes.

strontium A soft, malleable, ductile, metallic element.

stylet A relatively rigid elongated organ or appendage.

substrate A surface on which an organism grows or is attached.

substratum Any layer underlying the true soil.

succession A gradual process brought about by the change in the number of individuals of each species of a community and by the establishment of new species populations that may gradually replace the original inhabitants.

suctorian ciliates A subclass of ciliates in which adults are nonmotile, predatory, and unciliated and usually with suctorial tentacles for food capture.

sulfate-reducing bacteria Bacteria capable of reducing sulfate or other forms of oxidized sulfur to hydrogen sulfide.

sulfate reduction A process by which microorganisms reduce sulfate to organic sulfhydryl groups (R-SH).

sulfur oxidation Oxidation of elemental sulfur or hydrogen sulfide by microorganisms to hydrogen sulfate.

supernatant The liquid above settled solids.

surfactant A surface-active agent such as detergent that, when mixed with water, generally increases its cleaning ability, solubility, and penetration while reducing its surface tension.

suspension A mixture of fine, nonsettling particles of any solid within a liquid or gas.

symbiosis The living together of two dissimilar organisms, in which the association generally is advantageous or even necessary to one or both and is not harmful to either. Such a phenomenon is found among organisms in biological treatment processes.

synergistic An interaction between two entities producing an effect greater than a simple additive one.

synthesis Any process or reaction for building up a complex compound by the union of simpler compounds or elements.

tardigrades Any of a phylum (Tardigrada) of microscopic arthropods with four pairs of stout legs that usually live in water or damp moss. Also called waterbears.

taxonomy The study of the general principles of scientific classification.

telotroph An annelid larva having a preoral and posterior band of cilia.

teratogen An agent causing formation of a congenital anomaly.

terminal electron acceptor The final molecule that gets electrons during an oxidation–reduction reaction.

terrestrial Growing or living on land, as opposed to the aquatic environment.

testate Having or covered by a test (an external protective or skeletal covering), usually calcerous, silaceous, chitinous, fibrous, or membranous, secreted or built.

tetrad A group or arrangement of four.

thermophile Bacteria that grow best at temperatures between 45 and 60 °C.

thermophilic That group of bacteria that grow best within the temperature range of 45 to 60 °C.

thermotolerant Able to withstand high temperatures.

toxic Capable of causing an adverse effect on biological tissue following physical contact or absorption.

toxicity The property of being poisonous or causing an adverse effect on a living organism.

trace An extremely small but detectable quantity of a substance.

transmittance The fraction of radiant energy that, having entered a layer of absorbing matter, reaches its farther boundary.

treatability study A study in which a waste is subjected to a treatment process to determine whether it is amendable to treatment and to determine the treatment efficiency or optimal process conditions for treatment.

tricarboxylic acid cycle A sequence of enzymatic reactions involving oxidation of a two-carbon acetyl unit to carbon dioxide and water to produce energy for storage in the form of high-energy phosphate bonds. Also known as the Krebs cycle; citric acid cycle.

trichome A filamentous outgrowth.

trickling filter An aerobic, fixed-growth wastewater treatment process in which organic matter present in wastewater is degraded as it is distributed over a biological filter bed.

triglyceride A naturally occurring ester of normal, fatty acids and glycerol; used in the manufacture of edible oils, fats, and monoglycerides.

trinocular microscope A microscope with three ocular lenses, the third of which is used for placement of a photographic device.

trophic levels Any of the feeding levels through which the passage of energy through an ecosystem proceeds.

trophus Masticatory (chewing) apparatus in *Rotifera*.

trunk The main mass of the body, exclusive of the head, neck, and extremities; it is divided into thorax, abdomen, and pelvis.

tuberculosis A chronic infectious disease of humans and animals primarily involving the lungs, caused by the tubercle bacillus, *Mycobacterium tuberculosis*, or by *M. bovis*.

tungsten (1) Also known as wolfram. A metallic transition element, symbol W, atomic number 74, atomic weight 183.85; soluble in mixed nitric and hydrofluoric acids; melts at 3400 °C. (2) A hard, brittle, ductile, heavy gray-white metal used in the pure form chiefly for electrical purposes and with other substances in dentistry, pen points, X-ray-tube targets, phonograph needles, and high-speed tools and as a radioactive shield.

turbid Thick or opaque.

turbidity Suspended matter in water or wastewater that scatters or otherwise interferes with the passage of light through the water.

typhoid fever A highly infectious, septicemic disease of humans caused by *Salmonella typhi*, which enters the body by the oral route through ingestion of food or water contaminated by contact with fecal matter.

typhus Any of three louse-borne human diseases caused by *Rickettsia prowaszkii*, characterized by fever, stupor, headaches, and a dark-red rash.

ultraviolet radiation (UV) Light rays beyond the visible region in the visible spectrum; invisible to the human eye.

unsaturated Any chemical compound with more than one bond between adjacent atoms, usually carbon, and thus reactive toward the addition of other atoms at that point, for example, olefins, diolefins, and unsaturated fatty acids.

urea A natural product of protein metabolism found in urine; synthesized as white crystals or powder with a melting point of 132.7 °C; soluble in water, alcohol, and benzene; used as a fertilizer; in plastics, adhesives, and flame-proofing agents; and in medicine. Also known as carbamide.

vacuole A membrane-bound cavity within a cell; may function in digestion, storage, secretion, or excretion.

van der Waals force An attractive force between two atoms or nonpolar molecules, which arises because a fluctuating dipole moment in one molecule induces a dipole moment in the other, and the two dipole moments then interact.

vapor pressure For a liquid or solid, the pressure of the vapor in equilibrium with the liquid or solid.

vaporize To convert a liquid to a gas.

vegetative Of, relating to, or engaged in nutritive and growth functions.

ventral On or belonging to the lower or anterior surface of an animal, that is, on the side opposite the back.

virulence The disease-producing power of a microorganism; infectiousness.

virus The smallest (10 to 300 mm in diameter) life form capable of producing infection and diseases in humans and other animals.

viscosity The degree to which a fluid resists flow under an applied force.

volatile A substance that evaporates or vaporizes at a relatively low temperature.

volatilization The conversion of a chemical substance from a liquid or solid state to a gaseous or vapor state by the application of heat, the reduction of pressure, or a combination of these processes.

volute A twisted or spiral formation.

vulva The external genital organs of females.

waterbear A microscopic arthropod with four pairs of stout legs that lives in water or damp moss.

water hyacinth Floating aquatic plants whose roots provide a habitat for a diverse culture of aquatic organisms that metabolize organics in water.

wax Any group of substances resembling beeswax in appearance and character, and in general distinguished by their composition of esters and higher alcohols and by their freedom from fatty acids.

weirs A baffle over which water flows.

wetlands Surface areas, including swamps, marshes, and bogs, that are inundated or saturated by groundwater, often enough to support a prevalence of vegetation adapted to life in saturated-soil conditions.

wet mount Preparation of a microscope slide by placing a drop of sample water directly on the slide.

worst-case A testing situation in which the most unfavorable possible combination of circumstances is evaluated.

xenobiotic A completely synthetic chemical compound that does not naturally occur on earth.

yolk gland A modified part of the ovary that, in many flatworms and rotifers, produces yolk-filled cells serving to nourish true eggs.

zone settling When particle concentrations are sufficient enough so that particles interfere with the settling of other particles and they stick together.

zoogleal A jellylike matrix developed by certain bacteria. A major part of activated sludge floc and trickling filter slimes (zooglea).

Index

A

Acquired immune deficiency
  syndrome (AIDS), 7, 173,
  186, 189
Actinomycetes, 30, 37, 55, 146, 147,
  151–157
Activated-sludge process
  bacteria, 37, 55,
  filamentous organisms, 93–126,
    132–138, 140–156, 158
  flocculation, 55
  nematodes, 72, 74, 75, 81, 84, 88
  parasites, 195
  protozoa, 40, 45, 46, 49, 50, 51,
    53, 54, 58, 59
  rotifers, 64, 65, 66
  viruses, 86, 191
Adenovirus, 185
Aeration, 93, 103, 112, 135, 137,
  141, 142, 151, 152, 156, 157,
  158, 189
Aerobic bacteria, 35
Aerobic digestion, 33–38
Aerobic treatment, 40, 43, 44, 46, 50,
  54, 55, 65, 73–77, 144, 191
Aerosols, 181, 188, 189
Algae growth, 63, 65
Amoebae, 40–42, 45–48, 51, 55, 58,
  114, 115
Anaerobic bacteria, 35, 37
Anaerobic digestion, 35, 37, 38, 133,
  141, 144–146, 153, 157, 168,
  191, 192
Artificial support media, 58

Aspergillus, 180, 190
Astrovirus, 185
Aufwuchs community, 58
Autotrophic bacteria, 34, 37

B

Bacteria, 2, 3, 5–8, 23, 27–38, 45, 46,
  48, 54, 55, 58, 63, 65, 72, 73,
  77, 83, 84, 86, 94, 95, 115,
  117, 141, 146, 150, 151, 155,
  167–183, 188–193, 209, 210,
  221
Binocular microscope, 12, 13, 20, 21,
  104, 211
Bioaugmentation, 155
Biofilm, 37, 40, 48, 50, 54, 57, 58,
  75, 77, 195
Branching filaments, 30, 117, 121,
  122, 123, 124, 125, 129, 131,
  146, 147
Bristle worms, 7, 72, 79, 80
Burrow's polyvinyl alcohol (PVA)
  fixative, 202, 232

C

Campylobacter, 182, 183
Candida, 180, 190
Carbon dioxide, 28, 34, 35, 41, 45,
  50, 144
Cell wall, 30, 31, 32, 117, 118, 219

Chemical addition, filamentous bulking, 135, 137
Chemical addition, sulfide removal, 142
Chlorination, 74, 112, 115, 135, 140, 151, 154, 157, 158, 191, 192, 210
Ciliates, 6, 40–49, 50–55, 58, 66, 219
Ciliophora, 42, 43
Classification, 1, 2, 34, 40, 41, 74, 77
Clostridium, 34, 188, 189
Coagulation, 187, 190
Coliform bacteria, 167, 168, 171, 176, 191
Compost operations, 181
Compound microscope, 4, 12–15, 17, 20, 23, 212, 216, 217, 222
Counting chambers, 19, 58, 69, 77, 205, 225
Coxsackievirus, 185
Crystal violet sheath stain, 104, 118, 230
Crytosporidium, 186, 187
Cyclospora, 187
Cytoplasmic membrane, 31

D

Dark-field microscope, 21
Differential counts, rotifers, 69
Digononta, 61–66
Diluted sludge volume index procedure, 233
Diplogasteroidea, 74
Disinfection, 34, 186, 191, 192
DNA, 27, 34, 184

E

Echovirus, 185
Effluent limits, indicator organism, 171

Electron microscope, 24, 184, 185
Endospores, 34
Entamoeba histolytica, 186, 187, 200
Enterococci, 169–173
Enterovirus, 185, 191–193
Enumeration, 77, 169, 201, 202, 205, 225
Escherichia coli, 33, 55, 168, 169, 172, 182–185

F

Food-to-microorganism ratio (F:M), 103, 115, 132, 133, 135–137, 141, 143, 154, 158
Facultative anaerobic bacteria, 35, 37, 168
Fecal coliforms, 55, 167–171, 175, 176, 182
Fecal streptococci, 169, 170
Filament length, 99, 114, 121, 128
Filament measurement, 227
Filamentous bulking, 93, 96, 137, 138, 140, 145, 152
Fixative, 202
Flagella, 33, 41
Flagellates, 40, 41–48, 50, 51, 114, 115, 219
Floc formation, 54, 65, 95
Flotation analysis, 202
Flow rate, protozoa, 50, 51
Fluorescence microscopy, 24
Formalin, 67, 202, 233
Formalin–ether sedimentation, 204
Gas vacuoles, 34
Gastrotrichs, 72, 81–83
Giardia lamblia, 186, 187, 198
Glycocalyx, 32
Gram stain(ing), 31, 105, 119, 120, 126, 131, 221, 228

H

Hemacytometer, 19, 205, 225, 227
Hepatitis A virus, 181, 185, 192
Heterotrophic bacteria, 8, 34, 35, 39, 41, 45, 72, 188, 192
Human immunodeficiency virus (HIV), 7

I

Immunization, 181, 188
India ink, 102, 104, 112, 118, 140, 221, 230
Industrial wastes, 51, 180
Interference-contrast microscope, 23
Inverted microscope, 23

L

Lagoon, 37, 40, 41, 46, 48, 50, 53, 54, 64, 65, 75, 103, 183, 192
Leeches, 72, 79, 81
Legionella, 180, 188, 189
Leptospira, 183
Lugol's iodine stain, 205, 231

M

Mean cell residence time (MCRT), 103, 112, 115, 132–137, 141, 148–152
Membrane filter technique, 175, 176
Merthiolate–iodine–formalin (MIF) preservative, 202, 205, 232
Mesophilic bacteria, 35, 200
Metazoa, 5–8, 54, 71–88, 94, 102, 114
Methylene blue stain, 205, 231

Mixed liquor suspended solids (MLSS), 102, 135, 139, 140, 144, 148, 151, 158
Monogononta, 61, 62, 63, 64
Most probable number (MPN) technique, 175
Mycobacterium, 34, 147, 180, 188, 189, 191, 192

N

Neisser stain, 34, 103, 104, 106, 126, 127, 128, 129, 130, 131, 221, 229
Nematode, 2, 7, 8, 71–89, 114, 198, 200, 202
Nocardia, 30, 93, 100, 101, 113, 117, 129, 133, 134, 139, 140, 146, 147, 149, 151–155, 157
Norwalk virus, 186, 192

O

Oxygen consumption, 55, 77

P

Pathogen reduction, 190
pH, 50, 51, 65, 77, 133, 134, 135, 142, 145, 149, 150, 192
Phase-contrast microscope, 21, 23
Photomicrography, 104, 214, 215, 217, 218, 219, 221, 222
Pili, 33
Polarization microscope, 23
Poliovirus, 185, 193
Poly-β-hydroxybutyrate (PHB) stain, 34, 104, 230
Polymer addition, 155
Predation, protozoa, 6, 51, 54, 55
Preservatives and fixatives, 232

Process control, 52, 65
Protozoa, 2, 3, 5, 6, 8, 27, 28, 39–61,
    63, 72, 83, 84, 94, 102, 114,
    115, 167, 171, 179, 180,
    186–188, 191, 195, 197, 198,
    200, 202, 205
Psychrophilic bacteria, 35

Q

Quality assurance, 176, 177
Quality control, 176, 177

R

Receiving water criteria, indicator
    organism, 172
Rhabditoidae, 74
Rhizopoda, 41
Rotating biological contactor, 37, 40,
    52, 54, 57, 75
Rotavirus, 186, 187
Rotifer, 2, 6, 7, 61–70, 73, 79, 81,
    114, 115, 219
Roundworm, 7, 71, 73, 195, 198

S

Salinity, 171
Salmonella, 172, 180, 182, 183, 184,
    191, 192
Scanning electron microscope, 24
Scum disposal, 156
Scum formation, 100, 102, 147, 148,
    150, 154, 155, 156, 157
Sedimentation analysis, 202
Seed shrimp, 7, 72, 85
Sheather's sucrose solution, 203, 223

Shigella, 184, 191
Slime growth, 65
Slime layer, 65, 88
Sludge bulking, 93, 95, 96, 112, 135,
    136, 138, 140, 141, 142, 143,
    209
Sludge digestion, 35, 102
Sludge recirculation, 146
Sludge volume index (SVI), 95, 96,
    97, 98, 99, 100, 101, 113
Staining, 228
Stereoscopic binocular microscope,
    12, 13, 20, 21
Sucrose flotation, 202, 203
Sulfur oxidation, 104, 107, 118, 230

T

Taxonomy, 1, 34, 40, 67
Thermophilic bacteria, 35
Total coliform (TC), 168–172, 174
Total extended filament length
    (TEFL), 99, 113, 114
Toxic waste, 51
Transmission electron microscope,
    24
Trickling filter, 37, 40, 41, 46, 49,
    50, 51, 53, 54, 57, 65, 75, 77,
    103, 191, 195
Trinocular microscope, 211, 214,
    219, 220

U

Ultrasonication, 200

V

W

Y

Z

LaVergne, TN USA
12 May 2010

182362LV00003B/7/P